# SQUIRREL PROOFING your home & garden

RHONDA MASSINGHAM HART

*The mission of Storey Publishing is to serve our customers by publishing practical information that encourages personal independence in harmony with the environment.*

Edited by Dan Callahan and Jeanée Ledoux
Cover design by Meredith Maker
Cover photograph by © John C. Whyte, New England Stock Photo
Text design by Betty Kodela
Text production by Jennifer Jepson Smith
Line drawings by Jean Jenkins Loewer, and plans in chapter 7 by Brigita Fuhrmann
Indexed by Susan Olason, Indexes & Knowledge Maps

Squirrel Trail Mix on page 102 adapted, by permission, from the Birding New England Web site, http://www.tiac.net/users/pdragon/birdsqrl/html.

Bird feeder on page 143 designed by Sam Edson.



Printed in the United States by Versa Press
10 9 8 7 6 5

**Library of Congress Cataloging-in-Publication Data**

Hart, Rhonda Massingham, 1959–
Squirrel proofing your home and garden / Rhonda Massingham Hart.
p. cm.
Includes index.
ISBN 1-58017-191-5 (pb : alk. paper)
1. Ground squirrels — Control. 2. Wildlife pests — Control.
I. Title.
SB994.S67H27 1999
632'.69365—dc21 99-39634
CIP

# Contents

## *Dedication*

In memory of Billy the Squirrel,
who by now I can only guess has long since buried
his last pinecone. And in special admiration
for Billy's namesake, whom I have only once personally
witnessed up a tree,
William A. Forth, who is every bit as industrious,
discriminating, courageous, ambitious, resourceful,
and faithful as any squirrel. You're a good friend, Bill.
Don't let the nuts shake you out of your tree.

## *Acknowledgments*

A book like this is a wonderful opportunity to swap stories, both victories and defeats, regarding some of life's simplest pleasures and one of life's most fundamental frustrations — coping with wildlife on its own terms. Where sheer agility, wit, and will are concerned, squirrels are not to be outdone by mere human technology, a fact of which they constantly remind us.

I would like to thank the countless bird-feeding enthusiasts, gardeners, squirrel lovers, and others whose stories, good humor, and informed concern made this book possible. Special thanks to my children, Lance and Kailah, for putting up with the endless hours of Mom being chained to the computer. And my warmest thanks and appreciation to my sweetheart, friend, and counselor, Daymond Poe, for all his help while I was off in Squirrel Land. True friends understand when you go a little nuts.

And with great joy I thank the glory of God and Jesus Christ. They are what make all life's chores doable . . . and they brought us squirrels.

# *A Squirrel's-eye View*

**If it's any consolation,** you humans have it better than we squirrels.

For all the complaining and sputtering that you humans do regarding squirrels in *your* neighborhoods, perhaps you should stop to consider that we aren't always all that thrilled to have you in *ours*. After all, if we squirrels are messy or annoying with our chewing, digging, and helping ourselves to the wonderful buffets of bird feeders and garden goodies that you offer, humans are a real nuisance! You've leveled, rearranged, and in one form or another altered just about every part of our habitat. You are murder on trees, and you aren't that great at leaving things the way you found them.

However, humans have a positive, almost endearing, side as well. Some are downright cute, with shiny heads and really big eyes. Others are mildly amusing, coming and going the same routes day in and day out, nesting in the same cavities year after year, and entertaining us with some truly silly and bizarre mating behavior. Windows are one of your funniest adaptations.

But as humans go, we squirrels can tell you our favorites are the thinkers. Those persistent, clever individuals who attempt to solve problems with almost squirrel-like intelligence and diligence. Very impressive. Of course, they fail more often than not, but they are certainly fun to watch. It is on behalf of these humans, the thinkers, that we have banded together to approve the release of the information, much of it private and previously considered top secret, in this book. We hope that you will use it to form a better understanding of us. Follow the chapters on general squirrel habits and adaptations closely, then consider the sections, one at a time, that give you more insight into the separate species. You will learn much more about us and hopefully develop a true understanding of what we are all about. Turn, if you must, to chapters 5, 6, and 7 on how to "control" us (tee hee!), and ask yourselves if the continued violence against us is really necessary. Were things the other way around, would we not treat you much better?

Learn all you can about us and try to adopt our ways. Be thrifty, well prepared, and either social or not, depending upon your own nature. Laugh often, climb trees, and dig in the garden. All of this will make your lives that much better . . . once we take over the world.

CHAPTER 1

# Squirrels Are in Control

**Squirrels are in control,** but they don't want you to know it. They have orchestrated a huge, squirrelly conspiracy, cleverly designed to drive gardeners, bird-feeding enthusiasts, and anyone with an attic, nuts. When you stop to think about it, squirrels are everywhere, though you may not often see them. You're far more likely to hear their random chirps and chatter. Random, indeed! It's some sort of secret code. Ask anyone who's ever tried to foil the little sneaks. They are furry little secret agents and they know every move we make.

## *Where Squirrels Are Found*

They are everywhere. And they're watching you. Don't believe it? Leave a handful of sunflower seeds out on the lawn and count to 10.

Most of us fondly associate squirrels with oak trees and acorns. Weren't we all told as children that the continent owes its stately forests of oak and other hardwoods to the industrious squirrel? Those minions of the deep woods are credited with busily burying the nuts and seeds that ultimately sprout to life every spring to create a new generation of seedlings. If not for the squirrels, there might be no forests. Or so they would have you believe.

Many squirrels actually do populate the eastern forests. Personally, I believe they are operatives, planted there to make it look as though squirrels are the meek little woodland creatures they appear to be. They're a front. A deeper look into the secret lives of squirrels tells a much more ubiquitous story.

A quick glance at a population map shows that squirrels inhabit nearly every inch of North America. There is not a niche they haven't exploited. From the far Arctic reaches of northern Alaska and the Canadian territories to the arid deserts of Baja, California, and Mexico, squirrels are right at home. Further investigation proves that squirrels have conquered every sort of habitat, from the underworld to the very skies. Some tunnel through solid earth, while others glide on the breeze. Some sleep through the winter, while others cozy up with carefully laid-in stores to wait out the harshest weather. Squirrels have got it all figured out.

## *Population Shifts*

Given that we know they are behind every rock, tree, and shadow, it's easy to assume that squirrel populations have been constant through history, or perhaps even risen. Surprisingly, in some cases quite the contrary is true. Certain squirrel populations have fluctuated and floundered over the years to the very brink of disaster, due in no small part to human interference with habitat and food sources. Squirrel numbers are also known to fluctuate with nature's whims. A poor acorn crop one year may send tree squirrel populations plummeting, whereas a bumper chestnut crop, such as that in the George Washington and Jefferson National Forests in Virginia in 1997, often results in a sudden rise in numbers.

Take the case of the common gray squirrel. Early on, the squirrels did little to endear themselves to settlers. Hand-clearing ground was back-breaking work for those who sought to tame the wild land, but the squirrels saw it as simply a way to deprive them of their rightful homes. They retaliated, ravaging hard-earned crops with impunity. Exasperated farmers sought relief, often at the business end of a long-barreled rifle. Some states offered bounties for squirrel hides, and Ohio even allowed payment of taxes in squirrel "scalps," valued at three cents each. By the turn of the 20th century, conservationists feared the Eastern Gray Squirrel was headed toward extinction.

### Maybe It's the Nuts

Appropriately enough, the largest concentration of squirrels in the United States is in Washington, D.C., directly across from the White House in Lafayette Park. It figures: The capital of one of the leading industrial nations of the world is also the "Squirrel Capital" of the world. Wonder what attracts them all?

Some types of squirrels are endangered to this day. In many cases, individuals and governmental agencies are stepping in to ensure the squirrels get their fair shot at world domination. For instance, the state of Delaware is purchasing nearly 1,500 acres (6.1 square km) for the sole purpose of protecting its endangered Delmarva Fox Squirrel and the habitat it requires. Once found throughout the forests of the Delmarva Peninsula, which encompasses parts of Delaware, Maryland, and Virginia, the squirrel now occupies less than 10 percent of that original home turf, mostly hardwood and pine forests. The species has been listed as endangered since

## Conservation Counts

At the beginning of the 20th century, the population of the Eastern Gray Squirrel had been drastically depleted. Thanks to conservation efforts, today these squirrels are among the most common of wildlife.

1967, but is considered by experts to be more critically endangered today than when first listed. The most obvious reason is the loss of habitat as a result of deforestation and construction.

Another example of the efforts being made to save rare squirrels involves the $200 million Mount Graham International Observatory. This major telescope project in the Pinaleno Mountains of southeastern Arizona was halted in light of the habitat destruction that the project posed to the limited number of resident Mount Graham Red Squirrels *(Tamiasciurus hudsonicus grahemensis)*.

### Dogfights

It's not just tree squirrels that are under the gun. Their kin the Utah prairie dog was declared threatened in 1984, an upgrade from the endangered status it had held since 1973. Prairie dogs were most abundant around 1900 due to cattle management practices in their range that kept the surrounding grasslands at lower, preferred foraging height for the "dogs." Their delightful antics broke up the monotony for many a prairie traveler who watched from passing trains. One prairie dog town was reputed to cover some 25,000 square miles (64,883 square km), measuring 250 by 100 miles (403 by 161 km). But their heyday was short lived, because farmers and ranchers began to resent the competition. So prolific were the little "dogs" in their reproductive habits that a pair introduced to Nantucket Island as a novelty in 1890 had multiplied into a dangerous pest and nuisance, destroying crops and fields, by 1900. A town meeting called for the destruction of the critters, a task that was carried out in short order.

All-out war was declared throughout the West, and prairie dogs were destroyed en masse. The consequences were long standing. The packed soil of prairie dog ghost towns is poorly suited to growing forage, and once abandoned it takes many years to return to productivity. Also, the black-footed ferret was once thought to be extinct because of loss of the prairie dogs on which it depended for food and shelter. It remains a rare animal to this day.

**Squirrelly Facts**

The largest squirrel on earth is the Ratufa (*Ratufa indica*), or Indian Giant Squirrel, of Southeast Asia and Nepal. It can grow to a length of 3 feet (.92 m).

## *Happy Habitats*

Most of us tend to assume that where there's trees, there's squirrels. But when you consider the scope of squirreldom, you must contend with critters that tunnel, burrow, and soar, as well as climb. Squirrels: They're not just for trees anymore.

There is a squirrel for almost any imaginable habitat. Round-tailed Ground Squirrels *(Spermophilus tereticaudus)* and Antelope Ground Squirrels (*Ammospermophilus* spp.) thrive in the scorching, arid sands of desert regions, seeking shelter beneath windblown drifts that accumulate at the base of rocks, scrub bushes, creosote, mesquite, and other desert plants. Arctic Ground Squirrels *(Spermophilus parryii)* also inhabit sandy, barren tracts of land, but they prefer the extreme cold of the flat northern tundra. Chipmunks (*Tamias* spp.) routinely inhabit the deep cover of heavily forested areas as they burrow through the soft earth. Pygmy Squirrels (*Microsciurus* spp.) live in the Tropics, while the Kaibab Squirrel *(Sciurus aberti kaibabensis)* occupies only the isolated North Rim of the Grand Canyon. The Eastern

Fox Squirrel *(Sciurus niger)* exploits a range of habitats, including low-lying cypress and mangrove swamps, while both the Hoary Marmot *(Marmota caligata)* and the Columbian Ground Squirrel *(Spermophilus columbianus)* are at home in alpine meadows. Low, high, wet, dry, hot, cold — all told, squirrels are everywhere!

## A Niche of One's Own

Each squirrel subspecies has mastered its own place to call home. Some, however, are more versatile than others. For example, while some ground squirrels require arid, desolate homesteads, others, such as the infamous California Ground Squirrel *(Spermophilus beecheyi),* will take over any open area, from fields to rocky outcroppings to woodland hills. About the only similarity in preferred habitat for ground-dwelling squirrels is that they require soil with good drainage.

Tree squirrels have also laid claim to a range of habitats, from piney woods to hardwood forests to parks, cemeteries, and, of course, yards. The common denominator for the tree dwellers is a dependable supply of nuts or seeds. Trees also provide homes for these critters; especially popular are standing dead trees and those with hollow cavities. Fox squirrels tend to favor forests that are a mix of deciduous and coniferous trees bordered by agricultural areas. But when push comes to shove, they can be surprisingly adaptable. The Big Cypress Fox Squirrel, listed as threatened for more than 20 years, was originally indigenous to the Big Cypress Swamp area of southwestern Florida. These squirrels found new digs and became the subject of a study by wildlife biologists in 1995, when it was discovered that they prefer nesting in the swaying palms of luxurious golf courses to swamp living. Who can blame them? Sometimes these furry little land developers exhibit tastes strikingly similar to our own.

## You Can Run but You Can't Hide

Wherever you are, wherever you go, squirrels are sure to follow. While there are specific trees and plants you can use to purposefully attract more squirrels, little can be done to the natural environment to effectively deter them. Short of cutting down every tree and flooding every inch of soil, squirrels are here to stay. The catch is figuring out how to deal with them, and the first step is getting to know them.

> **Squirrelly Facts**
>
> The smallest squirrel in the world is the African Pygmy (*Myosciurus pumilio*), which grows to a length of only 2½ inches (6.4 cm).

# A Lesson on Squirreldom

**Think you know squirrels?** Don't be fooled into any sort of complacent belief that you know enough. That's just what they want you to think.

There are 273 species of squirrels worldwide, accounting for about 40 percent of all extant mammals. Sixty-six squirrel species are native to North America. They range in size from the fat and sassy woodchuck, which can top the scales at 15 pounds (6.8 kg), to the tiny Southern Flying Squirrel *(Glaucomys volans),* which weighs in at no more than 3 ounces (85 g). Some squirrels hibernate, and some don't; some estivate (sleep during hot weather), and some don't. Most are diurnal (active by day), while a few highly specialized, acrobatic species, the flying squirrels, roam the night skies. Not only are squirrels everywhere, but they work in shifts.

## *Squirrelly Attributes*

Squirrels have evolved some pretty nifty assets for dealing with their lot in life. First and foremost is attitude. Face it: If squirrels were the size of bears, not many among us would venture into their territory unarmed. But more than just a feisty demeanor empowers the mighty squirrel.

### Squirrel Talk

Squirrels of all types are wary and watchful, ever mindful of all that goes on around them, and constantly communicating their status to every other creature within earshot. Great communication systems keep them aware of their surroundings. From the whistles of prairie dogs and marmots to the chirps and chatter of tree squirrels, this is a talkative group. The normal frequency range is between .01 kilohertz (kHz) and 10 kHz. Such a wide range of sounds allows squirrels to communicate an equally diverse range of emotions, from a good laugh to anger to a warning of impending danger. The higher-pitched calls carry well, too: The shrill alarm call of the Northwest's Hoary Marmot is clearly audible for at least a mile.

Other creatures of the wild are highly attuned to the watchfulness of these ever present, self-appointed vigilantes of the woods. Deer will bound off at nothing more than the scolding chatter of a nearby squirrel. And animals as varied as birds flying overhead to predators on the ground will heed the squirrels' alarm as well.

#### Squirrel Medicine

Native Americans have long looked to the Medicine of the Squirrel for guidance. The squirrel is prepared for the future, quick on its feet, and ambitious. At a time when numerous prophecies foretell an age when all things will be in short supply, we would do well to follow the example set by the industrious squirrel.

Along with verbal expression, squirrels are very adept at certain forms of body language, often using their tails and gestures to communicate. For example, stamping the front feet is squirrel code for "Watch out!" A quick flick of the tail means, "Beat it, buddy! And don't you ever talk that way about my mother again!"

## Incredible Tails

Another key tool for most squirrels is that cute, fluffy tail. Who are the critters kidding? Do they think we don't know it's camouflage?

"Who, me?" innocent, liquid eyes seem to ask. "Why, I'm not up to a thing. Just sitting here idly beneath my nice, shady tail." Shady, maybe. Idly, never. It is from the feathery, plumed tail that the squirrel gets its name. Classified under the family Sciuridae, which translates directly to "shade tail," some species are deprived of the plumagelike tail, but those so equipped groom, pamper, and proudly display it with flair.

A squirrel's tail provides a handy umbrella during light rain showers and an equally useful parasol to shut out the hot sun. On cold nights, a squirrel's best friend is that warm, fluffy, built-in body warmer. Natural acrobats, tree squirrels rely on those plumed appendages as counterbalances for their constant high-wire high jinks. In fact, should a squirrel lose its grip, a spread tail acts as a sort of parachute to

The family name for squirrels translates to "shade tail"; their long, feathery tails protect them from the rain as well as the sun.

ease the fall. Should it land in water, the critter will use its tail like a rudder to swim to safety. When pursued, a squirrel will instinctively flick its tail erratically to confound its attacker, sometimes even losing a part of it to the onrushing predator. Hair, hide, or a few tail vertebrae are a small sacrifice to make for one's life. But a squirrel's best defense is not getting caught in the first place. Hidden in a blur of branches, twigs, and leaves, a fanned tail makes it all the harder to make out the familiar squirrel silhouette (see The Better Part of Valor, p.17).

## Armed and Dangerous

Grab on to any squirrel and two more of its customized features will become readily apparent. Squirrels are well armed, tooth and nail. Like other rodents, they have sharp — make that *sharp* — double incisors, and needlelike to talonlike claws.

The diet of most squirrels consists primarily of hard-shelled nuts and seeds, making a good set of choppers a must. Every squirrel is equipped with two large, curving sets of upper and lower incisors, as well as premolars and molars. The chisel-like incisors continue to grow throughout the squirrel's lifetime, averaging about 6 inches (15.2 cm) per year. The teeth can do the squirrel some serious damage if they are not constantly worn down by gnawing on nuts, bark, stems, electrical wiring, bird feeders, and unprotected human hands.

Squirrels chomp on wires' plastic coating because it cleans their teeth like dental floss.

Squirrel teeth provide some useful clues into the history of these renegade rodents. Through skull and dental comparisons, paleontologists have identified ground squirrel and chipmunk remains as old as 12 million years and

## Squirrel Floss

Ever wonder what maniacal plan prompts squirrels to relentlessly chew through anything within reach of those oversized choppers? A squirrel's teeth never stop growing. Unimpeded, they could easily reach 6 inches (15.2 cm) in length during its lifetime. They require constant use in order to remain sharp, clean, and whittled down to a usable length. Squirrels seem to especially like chewing on fibrous materials that can be pulled between the teeth, such as bark or the plastic coating on some kinds of wire. This shredding motion cleans the teeth in much the same way dental floss does.

tree squirrels as old as 28 million years. Interestingly, these were no "saber-toothed squirrels." Rather, the teeth of squirrels have changed little over the past 20 million years. Just think: The same teeth that protected the early emerging mammals from prehistoric predators are in use today against us.

Experienced squirrel fighters will tell you those teeth are serious weapons and can inflict a nasty bite. Many a spunky terrier has turned in terror after being perforated by squirrelly pearlies. The ground-dwelling squirrel species, from prairie dogs to marmots to small Antelope Ground Squirrels, use their teeth as shovel substitutes, slicing through soil, roots, and buried telephone cables as flying claws scurry to cast the debris aside.

As for those claws, they are a bit of a marvel as well. From the huge excavating extremities of marmots and ground squirrels, which can literally get the animals underground in as little as a minute, to the needle-fine, curved grappling hooks of tree squirrels, the appendages are a functional phenomenon. A tree squirrel can run up or down a tree trunk with equal aplomb, never faltering, never pausing to consider the possible consequences of a slip. Its ultrafine claws anchor the squirrel to nearly any surface, while those shiny, treacherous teeth grin tauntingly.

## Balancing Act

> **Squirrelly Facts**
>
> Squirrel skulls have been found containing teeth that have malformed and grown directly back into the jawbone, or curved upward toward the eye sockets.

If you have ever tried to shake a tree squirrel or chipmunk from a branch, pole, wire, rope, or other seemingly insecure perch, you were either ignorant of the varmint's uncanny sense of balance or just too mad to care. Merely watching from the ground for a few minutes can prove a dizzying experience. For tree squirrels, that fluffy tail is part of the secret to hanging in there, as are those super-grip claws, but a highly developed sense of balance is just one more attribute that makes the squirrel so versatile.

Though their acrobatic skills are exceptional, squirrels doing a high-wire act, such as crossing clotheslines or electric wires, are not invulnerable to the occasional misstep. This rarely results in a fall. Instead, as a squirrel tries to regain its balance, it treats viewers to a panicky but graceful display of swings, twists, and assorted gyrations. While a squirrel fumbling along a wire is amusing to us, it is humiliating to the creature itself. It ends the performance by retreating up the nearest tree and scolding the world.

A tree squirrel's anatomy is built for balance: The tail provides counterweight while the claws afford grip.

## Super Squirrel Strength

When you think of strength in animals, you probably conjure up images of powerful workhorses straining into their harnesses, or immense elephants moving heavy loads with their trunks, or perhaps even a hearty team of dogs pulling their master's sled across snow-encrusted trails. But how often have you said to yourself, "Dang, that squirrel is buff"?

Cute, fluffy, and strong: not an inaccurate description of the average squirrel. We owe much of our consternation over their antics to their physical prowess. Stop laughing and pay attention.

How high can you jump? Now imagine yourself a little over 14 inches (35.6 cm) tall. How high do think you could leap then? The average gray squirrel can easily hop 4 feet (1.2 m) into the air, and many can hit an 8-foot-high (2.4 m) target in a single bound. That's a leap of nearly seven times a squirrel's height. For most of us to jump as prolifically as a squirrel, we'd have to ascend nearly 40 feet (12.2 m). Impressed? Squirrels are right up there with bunnies and grasshoppers for powerful rear leg muscles that enable them to sprint onto bird feeders, eaves, troughs, fruit trees, and other places where they are not welcome.

Ground-dwelling squirrels also have their share of super-strength. Talented diggers, common woodchucks are the heavy equipment of the squirrel force. A single woodchuck typically moves some 700 pounds (317.5 kg) of soil in the excavation of its burrow system. The prairie dog, a cousin of the woodchuck, has also perfected the labor of burrowing and tunneling. Endowed with five (rather than most squirrels' four) front claws, prairie dogs have been known to move mountains — or at least rearrange a lot of flat land. Ground squirrels, too, are diggers par excellence. Even chipmunks are accomplished tunnelers, though you might never know it, since they are so adept at concealing the entrances to their hideouts.

## Alternate Transportation

Given that squirrels can walk, run, leap, and climb, one must at the very least appreciate their determination to get where they're going. But it's the extremes to which they push themselves that are truly amazing. Basic training for squirrels must be one tough boot camp.

Ground speeds have been unofficially clocked in a variety of strange instances. For one, almost any squirrel can outrun my dog, which is extremely unofficial and doesn't mean much, except perhaps for the fact that my dog can easily outrun me. One story that is still making the rounds concerns an Illinois state trooper who clocked a gray squirrel running across a highway with a radar gun. The squirrel wasn't pulled over, though — it was only doing about 20 miles (32.3 km) per hour, a pretty good clip for a critter with legs only a few inches long. Western Gray Squirrels are believed to be the fastest sprinters of the group. Alternately, fox squirrels have a top speed somewhere around 15 miles (24.2 km) per hour, pathetically slow for a squirrel. Marmots are even slower, clocking in at a pitiful 10 miles (16.1 km) per hour at a dead gallop. Squirrels of all types customarily zigzag, bound, and otherwise avoid dashing in a straight line, a clever strategy they have devised to help them elude predators.

Squirrels have also been documented swimming some quite impressive stretches of water. For example, gray squirrels crossed a ¼-mile (.4 km) span of the Mississippi River, between Wisconsin and Minnesota, in 1915. A 1930s account by W. J. Hamilton Jr. estimated that 1,000 squirrels

Cinderella's slippers were not originally made of glass. Instead, they were made of squirrel fur, which is eminently softer and more comfortable. A mistake was made in translating the story from the original folktale in the 1700s, and the poor girl has been dancing in glass slippers ever since.

swam down the Connecticut River all the way from Hartford to Essex, a journey of about 40 miles (64.5 km). Marmots seem to enjoy the water as well, often swimming about for a while in a lake or pond before exiting via the same bank from which they entered.

But climbing is the squirrel's true forte. Even the humble woodchuck can hoist itself upon a fence post or low-hanging tree limb, and a 'chuck eluding a hot pursuit can scramble a good 50 feet (15.3 m) up a tree. Rock Squirrels *(Spermophilus variegatus)* scurry up cliff faces that would make Sly Stallone nervous, and chipmunks zip up and down any obstacle in their paths. Indisputably, tree squirrels are the masters of the forest canopy, fearlessly leaping, scrambling, and, of course, climbing from tree to tree. In a densely forested area, aerodynamic tree squirrels can easily go about their day without ever lowering themselves to touch the ground, covering an area of well over a mile in many instances.

## In for a Dip?

Ground squirrels were routinely using my friend Lauren Allen's horse watering trough for a drink. They would climb a fence post to the top edge of the tank, leap onto the metal side, and dangle precariously toward the water to wet their whistles. Tired of fishing out drowned squirrels, Lauren came up with two solutions. One was to place a wooden plank inside the trough so the varmints could climb out. However, considering that horses tend to play with anything left in the water, she opted for plan two: heavy bricks stacked inside the trough leading up to one edge. Now, stuck squirrels can scramble to the brick "island" and bail themselves out of trouble.

In addition to the more conventional means of getting around, those sly squirrels have one more unexpected tactic. Some have flat-out shunned all that unnecessary climbing down tree trunks in favor of sailing. While it's true that what goes up must come down, graceful flying squirrels, or more accurately gliding squirrels, have simply found a better way. With a confident leap from a tree limb, they cut loose from the bounds of earth and soar to a nearby limb. They are the elite of the squirrel squadron, working only at night and effortlessly covering distances of up to 80 yards (73.1 m) in a single flight.

## The Better Part of Valor

Scare a squirrel and most likely its initial reaction will be to freeze. *Really* scare a squirrel and, depending upon its type and habitat, it will disappear in one of two ways: by diving for the cover of its burrow (or a nearby hole or rock crevice in the case of ground-dwelling squirrels), or fleeing up a tree.

In making a quick getaway, almost all types of squirrels will flick their tails, zigzag, or bound in seemingly random, peculiar movements — all this in an effort to confuse the pursuer. Once in the safety of a nearby tree, the squirrel will bolt to what it considers a safe height, and then do its best to keep the tree between itself and the danger lurking below. If avoiding an actual predator, the squirrel will press its body close to the tree bark and remain motionless. However, when a dog is the pursuer, the squirrel can't help but resort to one of its favorite activities: agitated, aggressive, consistent, verbal abuse. Chicka, chick, chick, chereeee!

## Territorialism

Many squirrels are notoriously territorial, which may have been the downfall of their scheme at world domination. The little buggers couldn't cooperate. But then we unwittingly helped them out. Habitat encroachment has forced more squirrels to live in closer proximity to one another, thus encouraging the very tolerance that now confounds so many hapless humans. The more squirrels trying to make a living in any one area, the more they must exploit whatever opportunities they can find, such as bird feeders, garden areas, and attic shelter.

In short, though most squirrel species are very aware of their property lines, they have learned to coexist, against their natural instincts, as an adaptation to the world we have created for them.

## *Squirrel Sense*

By now it should seem fairly obvious that squirrels have the advantage over just about any foe foolish enough to challenge them. So many specialized talents surely give them an edge. Nature has also blessed them with keenly adapted versions of the common senses we may take for granted.

### Vision

A squirrel's eyes are large and placed high on the sides of its head. This gives the ever watchful spy a wide field of vision, yet it never has to give away its position by so much as the slightest turn of its head. Constantly on the lookout, squirrels have excellent vision and are especially adept at noticing movement. Dim light does not impair their vision, and flying squirrels, with their large, orblike eyes, can see quite well in the dark.

### Scent

Perhaps the most stunning of a squirrel's senses is its sense of smell. In conjunction with a good memory (especially as compared to a human, who does well to remember where she left her car keys), the squirrel relies on its sniffer to find a winter stash or a potential

Squirrels depend on their keen sense of smell for finding food stashes beneath thick blankets of snow.

mate. One report states that a male squirrel can smell a female in estrus nearly a mile (1.6 km) away! Of the squirrels that cache or bury nuts and seeds, some have the habit of carefully burying them one at a time. Scientists who have observed squirrels handling each nut with care theorize that in so doing, they are marking them with their scent. (However, squirrels tend not to be terribly picky about whose nuts they unearth.)

Depending on memory to approximate the general location, squirrels must then sniff out the exact spot where each prize was buried. This is especially remarkable when snow cover further conceals the cache. The stately stands of oaks and other seed-grown trees throughout the American woodscape, many of them the result of forgotten winter stashes, illustrates that even squirrels aren't perfect.

## Hearing

Virtually all squirrel species rely on verbal communication, necessitating an acute sense of hearing. Chipmunks, for example, rely more heavily on their hearing than on their sense of smell, a rare trait in the world of rodents.

## Taste

When it comes to taste, squirrels are easily swayed by opportunity. Whatever is easiest to come by will quickly become squirrels' favorite food, hence their repeated ravaging of bird feeders and garden plots. Squirrels also have a lust for sweets, and they will hang upside down at hummingbird feeders to slurp the sweet, sticky liquid. Accordingly, fruits, berries, ripe corn, and tree sap are fair fare to almost any species of squirrel.

## Touch

Most squirrels have whiskers just above and below the eyes, on their throats, and on their noses. These whiskers, or vibrissae, act as touch receptors, keeping a squirrel constantly apprised of its immediate surroundings. The whiskers are a tremendous aid during breakneck runs through the trees.

Cute and cuddly looking, squirrels are not especially tactile with one another except when raising their young. Male squirrels love to preen and will take as much as twice the time females do to groom themselves. I imagine it simply feels good.

"The wildness of squirrels is an awesome wildness."
– Douglas Fairbairn, *A Squirrel of One's Own*

CHAPTER 3

# Identifying the Enemy

**It's about time somebody told you the truth about squirrels.** Though some do a good impression of a cute little woodland creature, others don't mind throwing their weight around. I'll discuss six types of squirrels here, including some lumbering relatives you may not think of as squirrels. Marmots and prairie dogs are the oddball second cousins of ground squirrels, tree squirrels, and flying squirrels, as are those striped bandits, the chipmunks. All have in common basic distinct characteristics, from similar skull and teeth formation to a penchant for devising ways to drive humans nuts. Here they are, exposed for what they really are.

**Note about the maps**: Squirrels and their relatives inhabit many parts of the globe, but only their North American ranges are illustrated here.

## *Marmot* (*Marmota* spp.)

Typically big, fat, slow squirrels that can't climb very well, these lumbering varmints are also referred to as woodchucks and groundhogs *(Marmota monax)*. Still don't believe that squirrels have conquered the earth and are in cahoots against us? Consider the case of Punxsutawney Phil, the famous groundhog that predicts the coming of spring. He even has his own holiday, for Pete's sake!

And just how is Phil able to tell so much merely by popping his head out of his burrow for a moment? The woodchuck's eyes, ears, and nose are all set high on its head, which allows it to stay nearly submerged below ground level while still maintaining surveillance on its surroundings (along the same lines as a periscope on a submarine). Perhaps somewhere in its wistful gaze it detects the subtle changes of spring already in process, or the clutches of winter still firmly grasping the land. Perhaps the linger of frost on its whiskers is a telltale indicator to the wizened rodent. Or maybe it's got some complex weather-prediction computer system in its den.

### Who Lives in That Burrow?

When a woodchuck is in residence, you'll see a mound of soil at the main entrance. When the woodchuck vacates and another boarder moves in, the mound is worn away. A little espionage should reveal scat, bones, or other debris around the den, depending on the new resident. Knowing what is living where is an advantage in preventing animal damage to your yard and garden.

Aside from the woodchuck, there are four more species of marmots in North America:

- The Yellow-bellied Marmot *(Marmota flaviventris)* in the West
- The Hoary Marmot *(M. caligata)* in British Columbia and Alaska
- The Olympic Mountain Marmot *(M. olympus)*
- The Vancouver Island Marmot *(M. vancouverensis)*

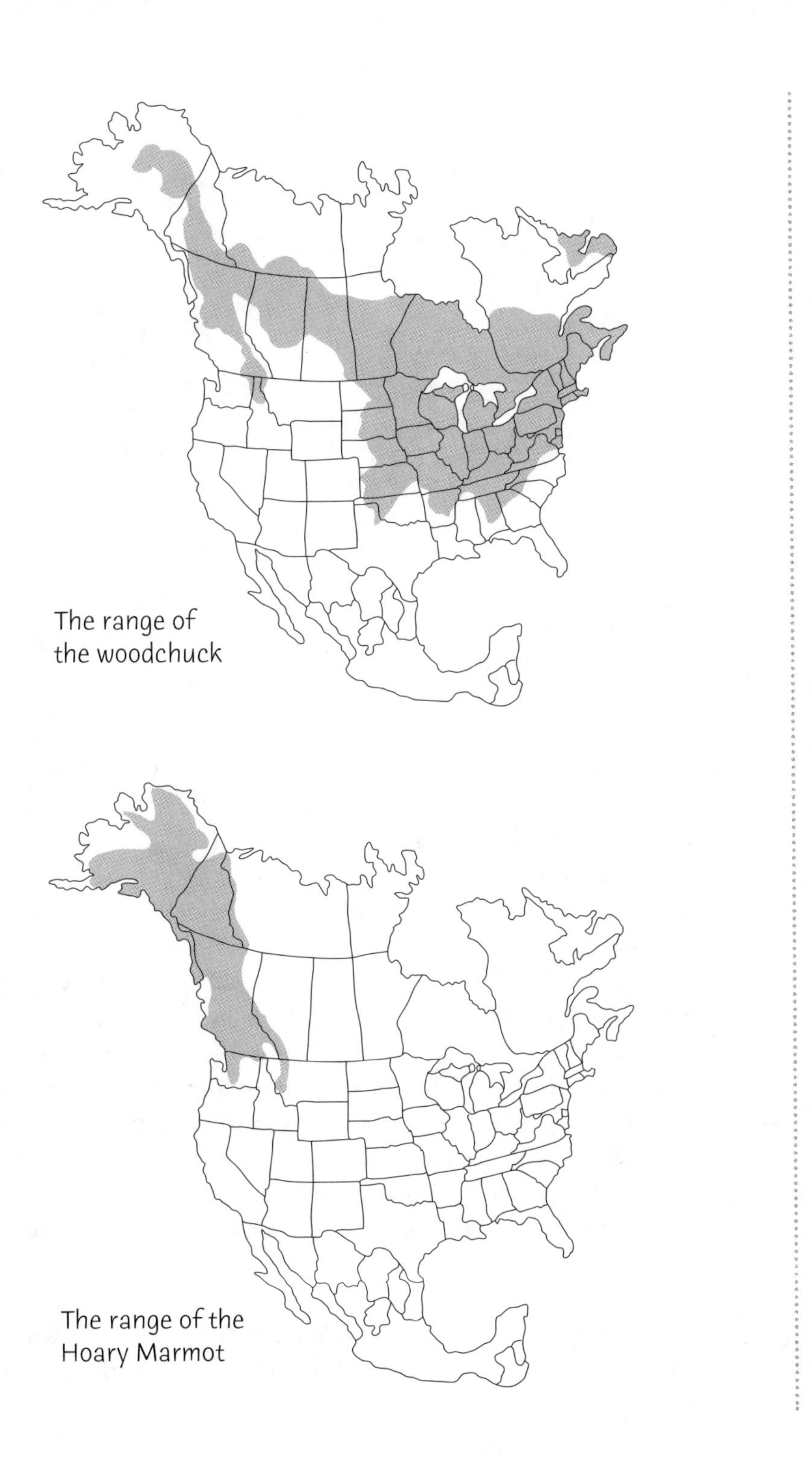

The range of the woodchuck

The range of the Hoary Marmot

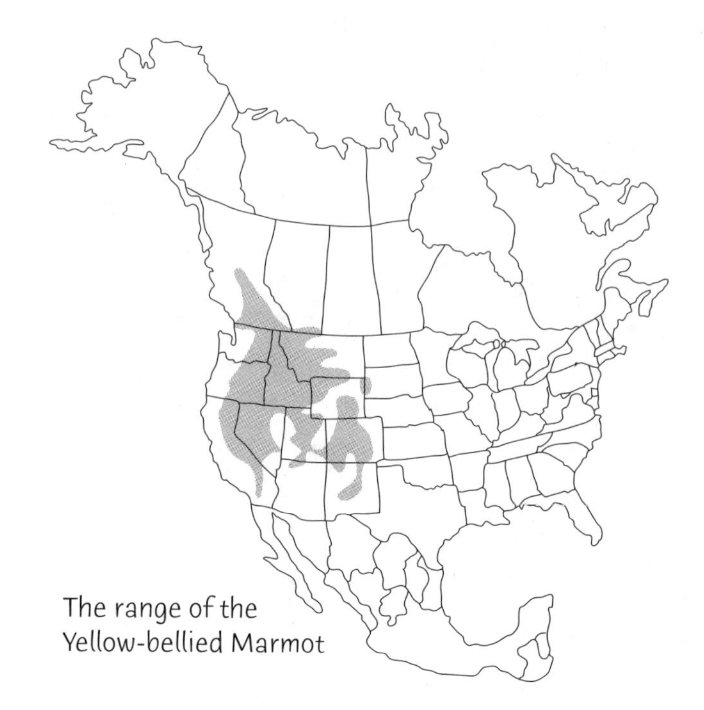
The range of the Yellow-bellied Marmot

Most marmots are more interested in each other than in whatever you may be up to, and they usually mind their own business, spending their days sunning themselves, resting in their burrows, and foraging for food in their respective wildlife habitats. It's those hermits, the woodchucks, you have to watch out for.

Woodchucks are chunky critters that weigh between 4 and 15 pounds (1.8–6.8 kg). They have long, coarse, brownish fur, four industrial-strength claws on each front foot, short tails, and cute little ears. They usually waddle along in a ground-hugging walk and can often be spotted sunning themselves on some elevated vantage point, such as a mound of earth or a fence post. They are found throughout the eastern United States, except in the Deep South, and their relatives inhabit much of the Upper Northwest.

The average woodchuck can decimate a home garden in a couple of good forays. A pair can ruin weeks of hard work, tender care, and careful planning virtually overnight. Each consumes more than a pound (.45 kg) of vegetable material per day, and much of what they don't consume gets trampled beneath their pudgy little feet. They have a taste for soybeans, corn, alfalfa, peas, beans, carrot tops, lettuce, and squash. They have even been known to gobble up a garnish of several types of flowers. But what they *crave* is dessert.

### Squirrelly Facts

A squirrel's teeth continue to grow throughout its lifetime.

Like most squirrels, woodchucks will go to extremes to satisfy their sweet tooth. Normally, they are rather homebound individuals, not foraging more than 50 yards (45.7 m) from their burrow entrance. But when melons are ripe, or apples or other tree fruits are falling, woodchucks will venture far and wide for such sweet rewards.

The clever complexity of the woodchuck den system is another tip-off that these squirrels are a whole lot smarter than we've been led to believe. In order to establish any significant type of control program, it helps to understand the burrow systems and how woodchucks use them. They typically construct both winter quarters and at least one nice little summer place. Digging out a second summer burrow allows the resident to forage farther afield while staying within range of a hideaway. Winter quarters differ from summer quarters in that they are basically extravagant bedrooms, often strategically situated beneath a stump, tree, or debris pile. The tunnels

The woodchuck or groundhog — subject of tongue twisters and folklore — is a common North American marmot.

## Marmot Stats

**Average life span**: 4–6 years
**Average size**: Length 18–27 inches (45.7–68.6 cm); weight 4–15 pounds (1.8–6.8 kg)
**Positive ID**: Small ears, short legs, brown fur, bushy tail; mounded burrow entrances
**Hibernation**: Mid-October through February
**Habitat**: Pastures, meadows, suburban yards, parks, gardens
**Range**: Much of Canada and the eastern United States
**Signs**: Large burrow entrances, up to a foot in diameter, with mounds of dirt near the opening; often there are extra openings nearby, without the dirt mounds; pathways leading to and from the burrows may be trampled down in tall grass
**Territory**: Approximately 100–200 yards (91.4–182.8 km) in diameter
**Number of young born per year**: 4–6, in March or April
**Most-active periods**: Warmer parts of early spring days, cooler times of summer days
**Diet**: Green vegetation, grass, clover, plantain, corn
**Natural enemies**: Parasites, red foxes, coyotes, badgers, and people
**Positive contributions**: Fertilization and aeration of soil

usually run more than 25 feet (7.6 m) in length, at a depth of 2 to 4 feet (.61–1.2 m). The boudoir is frequently placed discreetly at the far end. Preferring to live alone and only occasionally tolerating a roommate or two, woodchucks spend most of their day snoozing in the burrow. They are most active in the early morning and again in the early evening, the best times to be on guard. On wet, chilly days, they tend to stay at home. After a woodchuck moves on, its burrow becomes an important habitat for other wildlife, such as rabbits, skunks, foxes, and badgers.

Knowing when woodchucks are out and about should help your efforts to stymie their raids. They are true hibernators. After thoroughly plundering your fruit crop, they store up fat and settle down for a long winter's nap in mid-October, from which they awaken in February. So the garden is safest from late fall through the dead of winter — quite the comfort, eh? They either see their shadows or they don't, but soon after they find one another, they mate (about the only time woodchucks seem to want anything to do with one another), producing litters of four to six babies about a month later. By early July, the kids are on their own. And that's when the real trouble starts. Hungry, unsupervised teenage squirrels are on the loose.

But before you round up a small army of neighbors, pitchforks at the ready, consider the fact that woodchucks actually contribute to their environment (your yard) in a positive way. Each burrow has a separate "bathroom," which the woodchucks use fastidiously. These chambers help to fertilize the soil. Their infernal burrowing keeps the soil loose and aerated, mixes in lots of water and organic material, and brings up subsoil — a bona fide boon to soil health.

## *Prairie Dog*
### (*Cynomys* spp.)

A uniquely American creature, the prairie dog was known only to Native Americans, French trappers, and buffalo until its discovery by the Lewis and Clark expedition in 1804. By the time of its introduction to the scientific community, the prairie dog had already earned several descriptive nicknames, including barking squirrel, prairie squirrel, prairiehunde, and petit chien. First considered to be a form of marmot (and so dubbed *Arctomy ludovicianus* by George Ord of the

Prairie dogs build miles of underground tunnels that, like our cities, are divided into unique neighborhoods.

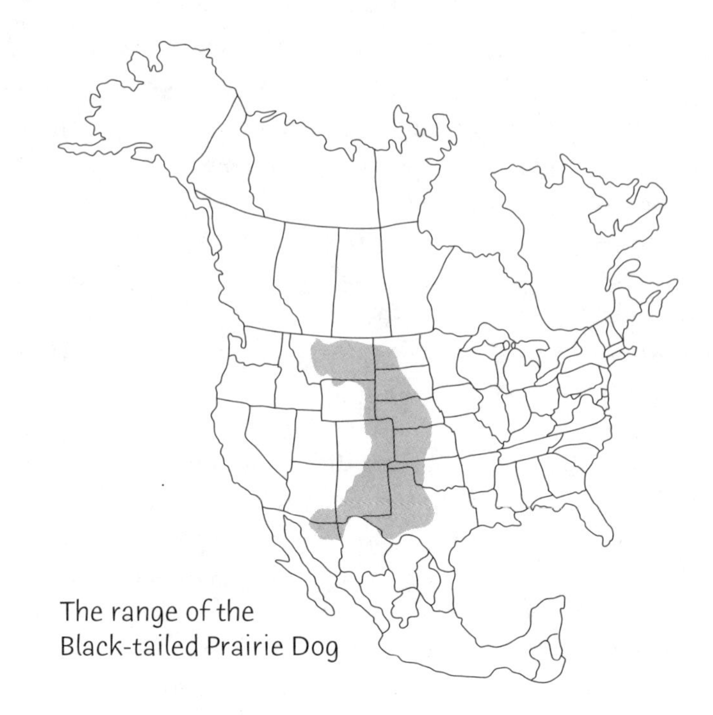
The range of the Black-tailed Prairie Dog

Lewis and Clark expedition in 1815), it was soon accorded its rightful individuality and renamed *Cynomys,* or "doglike mouse." There are several prairie dog varieties, from the endangered Black-tailed Prairie Dog, the largest of the group, to the smaller White-tailed Prairie Dog, Gunnison's Prairie Dog, and Utah Prairie Dog.

Prairie dogs are the masters of colonization, the extreme opposite of the solitary woodchuck. Even before Manifest Destiny overran the western half of the New World, prairie dogs had been there, done that, and taken to the underground. Vast systems of tunnels cover hundreds of sites, from Saskatchewan to Mexico. The colonies, or towns, often spread under miles of ground and, much like human cities, are divided into districts, or suburbs, called *coteries.* Each coterie covers about ¾ acre (3035 square meters). Members of the same coterie share tunnels and greet each other with sniffs and prairie dog "kisses," a form of recognition, not flirtation. Rivalries between coteries are fierce, and prairie dogs from neighboring coteries will drive each other away if they stray into the wrong territory. But despite this contentious behavior, all the dogs of the community are in sync when it comes to pulling guard duty. Every little set of eyes watches for the shadows of eagles, hawks, owls, or any suspicious movement on the ground. All ears tune in. At the slightest hint of danger, the alert is sounded with a sharp whistle; all in earshot beat a hasty retreat to the safety of the vast underground network of tunnels.

Before you smugly assure yourself that you are safe from the gnawings of these oversized ground squirrels, you should know that prairie dog towns are continually expanding underground. Generation after generation, constant tunneling and expanding about the edges of the towns serves to gradually move the animals along the prairies, conceivably in search of new forage and fresh ground. The towns migrate in vast, slow-motion waves across huge stretches of open grassland. Although it is estimated that roughly 90 percent of the animals' original habitat has been lost, prairie dog towns today are much the same as they were in Lewis and Clark's time, though notably smaller.

## A New Discovery

"We arrived at a spot . . . nearly four acres in extent . . . covered with small holes. These are the residence of a little animal, called by the French petit chien (little dog), which sit erect near the mouth and make a whistling noise, but when alarmed take refuge in their holes. . . . The petit chiens are justly named, as they resemble a small dog in some particulars though they have also some points of similarity to the squirrel. The head resembles the squirrel in every respect except that the ear is shorter; the tail is like that of the ground-squirrel; the toenails are long, the fur is fine and the long hair is gray."

— Capt. Meriwether Lewis, September 7, 1804, from *The History of the Lewis and Clark Expedition*

## Prairie Dog Stats

**Average life span**: 4–5 years
**Average size**: Length 12–16.5 inches (30.5–41.9 cm); weight 1.4–3 pounds (.64–1.4 kg)
**Positive ID**: Yellowish tan to brown with dark backs, light bellies, and either white- or black-tipped stubby tails, depending on species; some are stocky, others more streamlined; small, perky ears; large dark or black eyes
**Hibernation**: None
**Habitat**: Prairie land
**Range**: Saskatchewan to Mexico
**Territory**: Towns of a few acres to 100 acres (28.7 square km) and up
**Number of young born per year**: 1–8, in April or May
**Most-active periods**: Mornings and early evenings
**Diet**: Green plants, grass, occasional grasshoppers
**Natural enemies**: Black-footed ferrets, foxes, badgers, coyotes, eagles, hawks, and snakes
**Positive contributions**: Food source for predators, amusement to train travelers, dust wallows for buffalo

Prairie dogs are smaller than marmots, but stout, short of limb and tail, and as full of attitude as any member of the squirrel family. They average between 12 and 16.5 inches (30.5–41.9 cm) in length and weigh in at 1.4 to 3 pounds (.64–1.4 kg). Their dust-colored fur lies close to the skin. The most you are likely to see of any given individual is its head and upper body as the prairie dog pokes its nose out of a tunnel opening. However, they are easily identifiable by their surroundings. If you are in the middle of a vast stretch of holes, and a squirrel-like critter pops its head out of the ground and whistles at you, it's not a compliment; it's a prairie dog.

Even though prairie dog numbers have sharply declined, the animals are still considered a nuisance in some areas of the American West. Many ranchers believe that prairie dogs compete with cattle for grass. They do create what some would consider an unsightly mess, with holes that can be dangerous to livestock, and their appetite for grass is legendary. But studies have shown that they do not pose any real threat to grazing lands. Still, any folks living within the vicinity of a prairie dog town might well fear for their lawns.

## *Ground Squirrel*
### (*Spermophilus* spp.)

Characterized by the genus name *Spermophilus,* which means "seed lover," ground squirrels are often misidentified as either gophers or, in the case of the Golden-mantled or Thirteen-lined Ground Squirrel, chipmunks. Considering their well-earned reputation as troublemakers, those disguises might be a good idea.

The range of the
Arctic Ground Squirrel

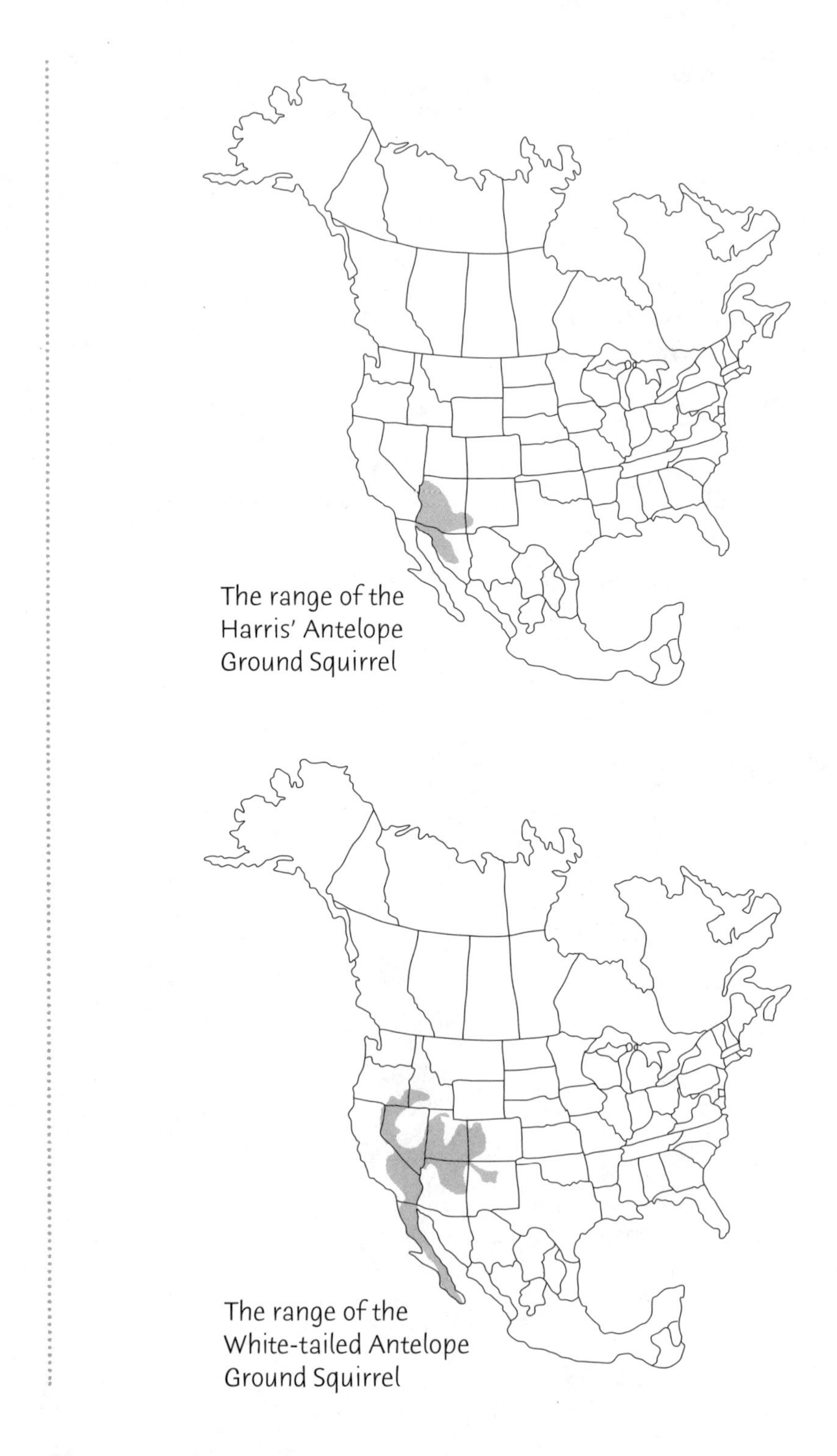

The range of the Harris' Antelope Ground Squirrel

The range of the White-tailed Antelope Ground Squirrel

## The Many Different Ground Squirrels

Their predators will tell you there are 19 flavors, or different species, of ground squirrels in North America, some with very different habits from the others and only a couple of which give the whole lot a bad name in gardening circles. The tiny antelope squirrels (genus *Ammospermophilus*) are true ground squirrels that spend their days quietly munching cactus seeds and insects in the heat of America's southwestern deserts. Likewise, the oversized Arctic Ground Squirrel *(Spermophilus parryii),* whose life on the tundra threatens nary a tulip nor tuber, poses no great worry to gardeners. It's their California cousins and a few eastern relatives you have to watch out for.

## Common Characteristics

The many varieties of ground squirrels have much in common. They live to dig, sleep, eat, and mate. Most often they don't disperse their young far from the home burrow system, and they live in close proximity to one another (despite often not being all that fond of one another), in some cases with as many as five thousand individuals per square mile (2.6 square km). Though their burrow systems are often extensive, sheltering varying numbers of squirrels, they tend to bicker or avoid each other the majority of the time.

A typical day primarily consists of snoozing underground, burning off fat. It's how they put on that fat that puts them in contention with most gardeners. Most ground

### Telltale Tails

Antelope squirrels are most readily distinguished from other types of ground squirrels by the way they hold their tails when scurrying away. Like its namesake, the pronghorn antelope, the antelope squirrel runs with its tail held vertically at full mast, exposing its white fanny, while other ground squirrels run with their fluffy tails trailing horizontally behind them.

squirrels have a diet of roughly 50 to 80 percent vegetation. One study, which collected nearly 1,200 Columbian Ground Squirrels from a 200-acre (.81 square km) tract of land, concluded that they consumed enough vegetation to support 12 sheep or 3 cows. Stuffing fur-lined cheek pouches with seeds, ground squirrels can haul an amazing amount of material back to the nest. They indulge in a variety of roots, bulbs, grasses, and flowers, too. Perhaps the only saving grace of the pests is that they also consume heaping quantities of weed seeds and insects. The Thirteen-lined Ground Squirrel devours at least half its diet in the form of grasshoppers, crickets, caterpillars, beetles, ants, and insect eggs.

Like other squirrels, ground squirrels keep a watchful eye on their surroundings and are ever ready to sound the alert. Danger may come from above, in the form of eagles, hawks, and falcons; from the ground, as foxes, coyotes, badgers, and bobcats prey on them; from below ground, because they are favorite snake snacks; and from us, as we attempt to preserve our yards and some fraction of our sanity.

The range of the California Ground Squirrel

### California Ground Squirrel

Understandably, there is no fan club for California Ground Squirrels *(Spermophilus beecheyi),* which are generally considered pests and layabouts. They are small (less than 2 pounds [.91 kg]) brown squirrels with unusually bushy tails for ground dwellers. A dark or black V on the back, pointing toward the nose sets them apart. Most of their

## California Ground Squirrel Stats

**Average life span**: 2–3 years
**Average size**: Length 9–11 inches (22.9–28 cm); weight 1.5–2 pounds (.68–.91 kg)
**Positive ID**: Brownish gray with a dark or black V on the back
**Hibernation**: Varies: Some ground squirrels sleep from mid-October through April; others split the season, estivating during the heat of summer and hibernating through the worst of winter
**Habitat**: Pastures, meadows, suburban yards, parks, gardens
**Range**: Western coast of the United States and Mexico
**Signs**: Burrow entrances, sometimes with narrow pathways trampled in the grass
**Territory**: Usually about 50–100 yards (45.7–91.4 m) from the burrow opening
**Number of young born per year**: Usually 1 litter in spring, averaging 3–8 offspring
**Most-active periods**: Daylight hours during spring and early fall
**Diet**: Grass, roots, bulbs, stems, leaves, insects, and, in some cases, eggs and smaller mammals
**Natural enemies**: Birds of prey, foxes, coyotes, badgers, bobcats, snakes, and people
**Positive contributions**: Fertilization and aeration of soil; consumption of harmful insects and weed seeds

waking hours are spent eating, often damaging crops or garden plants, or digging holes that frequently trip up people and livestock. Several of these rodents coexist, antagonistically as it may be, in one burrow system, each squirrel using its own private entrance. One group of researchers discovered a system of tunnels that housed five males and six females in 741 feet (226 m) of tunnel space, with 33 separate entrances.

Though they tend not to venture more than 100 feet (30.5 m) or so from their underground burrows, this offers little comfort to the gardener, as the squirrels' ever expanding burrows are constantly under construction. Major renovations take place in spring, just as you are trying to coax tender new plants along. California Ground Squirrels spend a good part of each day above ground during the February-to-April breeding season, but they spend an inordinate amount of time sleeping in their dens, hibernating for several months at a time.

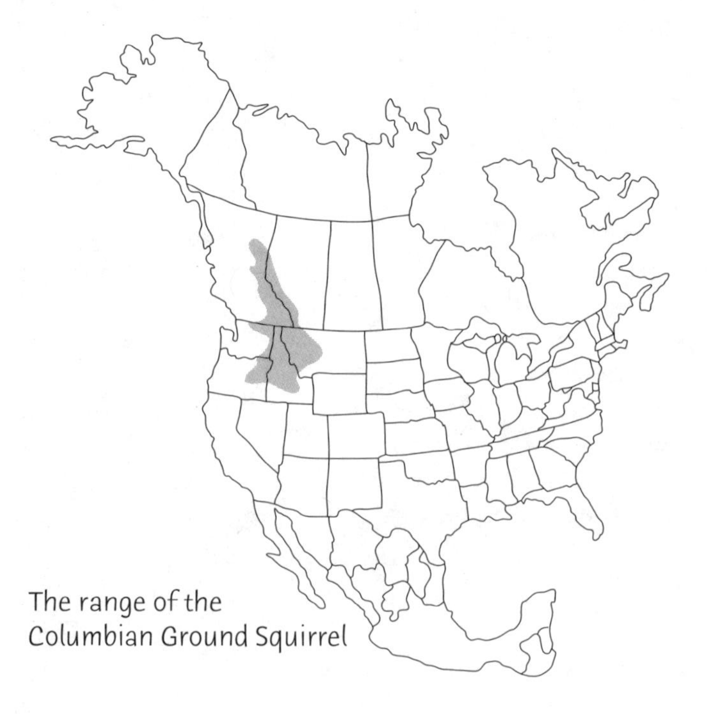

The range of the
Columbian Ground Squirrel

## Columbian Ground Squirrel

Slightly to the east, another bushy-tailed but not-so-bright-eyed relative, the Columbian Ground Squirrel *(Spermophilus columbianus)*, enjoys a similar lifestyle, snoozing away as much as seven or eight months of the year in specially prepared hibernation dens. These small grayish to buff-colored squirrels generally prefer open grassy areas, but they also move into wooded areas, housing developments, and parks.

### Columbian Ground Squirrel Stats

**Average life span**: 2–3 years
**Average size**: Length 8–11 inches (20.3–28 cm); weight 1.5–2 pounds (.68–.91 kg)
**Positive ID**: Grayish to buff-colored fur
**Hibernation**: Varies: Some ground squirrels sleep from mid-October through April; others split the season, estivating during the heat of summer and hibernating through the worst of winter
**Habitat**: Pastures, meadows, suburban yards, parks, gardens
**Range**: Northwestern United States and southwestern Canada
**Signs**: Burrow entrances, sometimes with narrow pathways trampled in the grass
**Territory**: Usually about 50–100 yards (45.7–91.4 m) from the burrow opening
**Number of young born per year**: Usually 1 litter in spring, averaging 3–8 offspring
**Most-active periods**: Daylight hours during spring and early fall
**Diet**: Grass, roots, bulbs, stems, leaves, insects, and, in some cases, eggs and smaller mammals
**Natural enemies**: Birds of prey, foxes, coyotes, badgers, bobcats, snakes, and people
**Positive contributions**: Fertilization and aeration of soil; consumption of harmful insects and weed seeds

## Franklin's Ground Squirrel

Farther east, you might encounter the Franklin's Ground Squirrel *(Spermophilus franklinii)*, a big, nasty-tempered, dark-colored character that spends up to 90 percent of its time underground. Put several of these squirrels together and they will fight among themselves. They also threaten local wild bird populations with their taste for duck and pheasant eggs. Though it might endear them to some that they tend to avoid manicured lawns and fields in favor of brushy banks, they occasionally pop up underneath compost piles, putting nearby greenery in peril.

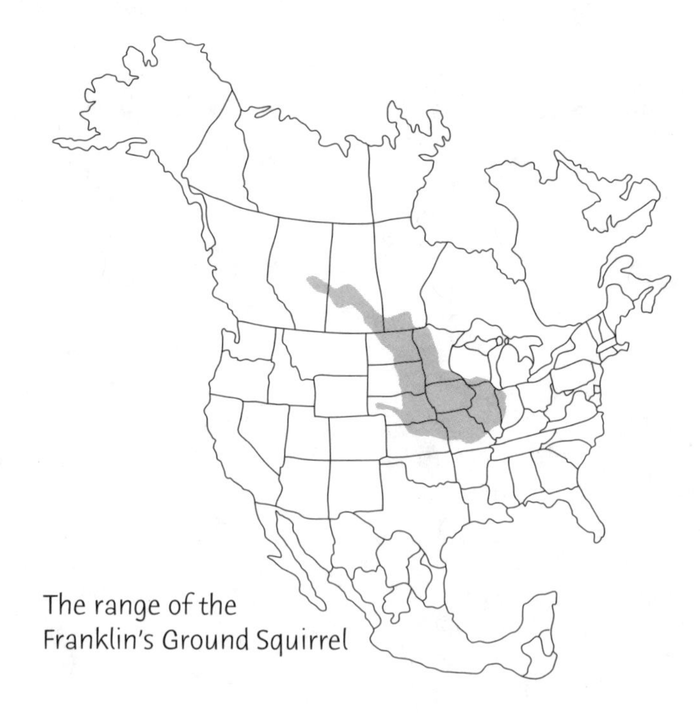

The range of the Franklin's Ground Squirrel

## Franklin's Ground Squirrel Stats

**Average life span**: 2–3 years
**Average size**: More than 15½ inches (39.4 cm) in length and 1¾ pounds (.8 kg)
**Positive ID**: Large and bad-tempered with dark-colored fur
**Hibernation**: Varies: Some ground squirrels sleep from mid-October through April; others split the season, estivating during the heat of summer and hibernating through the worst of winter
**Habitat**: Pastures, meadows, suburban yards, parks, gardens
**Range**: Midwestern United States and southern Canada
**Signs**: Burrow entrances, sometimes with narrow pathways trampled in the grass
**Territory**: Usually about 50–100 yards (45.7–91.4 m) from the burrow opening.
**Number of young born per year**: Usually 1 litter in spring, averaging 3–8 offspring
**Most-active periods**: Daylight hours during spring and early fall
**Diet**: Grass, roots, bulbs, stems, leaves, insects, and, in some cases, eggs and smaller mammals
**Natural enemies**: Birds of prey, foxes, coyotes, badgers, bobcats, snakes, and people
**Positive contributions**: Fertilization and aeration of soil; consumption of harmful insects and weed seeds

## Too Much of a Good Thing?

Richardson's Ground Squirrels were once so plentiful that the eminent writer and editor of the Lewis and Clark journals, Elliott Coues, wrote, "Millions of acres are honeycombed with its burrows. . . . I never saw any animals — not even Buffalo — in such profusion."

## Richardson's Ground Squirrel

The small yellow-gray Richardson's Ground Squirrel *(Spermophilus richardsonii),* or picket pin, occupies a relatively small range in the western United States and southern parts of Canada, but its impact has been substantial. The greatest threat these critters pose to human society may well be their sheer numbers and potential for uncurbed reproduction. The many dens of this species were thought to have once rivaled the multitudes of prairie dog mounds in number, stretching for miles on end across the midwestern plains.

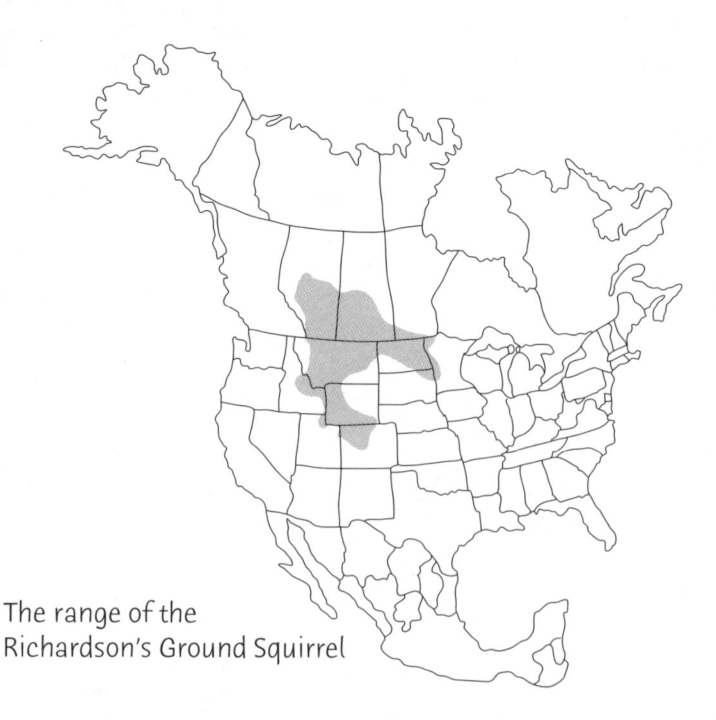

The range of the Richardson's Ground Squirrel

## Richardson's Ground Squirrel Stats

**Average life span**: 2–3 years
**Average size**: Length about 8 inches (20.3 cm); weight approximately 1.5 pounds (.68 kg)
**Positive ID**: Small with yellow-gray fur
**Hibernation**: Varies: Some ground squirrels sleep from mid-October through April; others split the season, estivating during the heat of summer and hibernating through the worst of winter
**Habitat**: Pastures, meadows, suburban yards, parks, gardens
**Range**: Western United States and southern Canada
**Signs**: Burrow entrances, sometimes with narrow pathways trampled in the grass
**Territory**: Usually about 50–100 yards (45.7–91.4 m) from the burrow opening
**Number of young born per year**: Usually 1 litter in spring, averaging 3–8 offspring
**Most-active periods**: Daylight hours during spring and early fall
**Diet**: Grass, roots, bulbs, stems, leaves, insects, and, in some cases, eggs and smaller mammals
**Natural enemies**: Birds of prey, foxes, coyotes, badgers, bobcats, snakes, and people
**Positive contributions**: Fertilization and aeration of soil; consumption of harmful insects and weed seeds

## Thirteen-lined Ground Squirrel

Finally, the cleverly disguised Thirteen-lined Ground Squirrel *(Spermophilus tridecemlineatus)* rates mention for the vast territory this chipmunk look-alike occupies. Nicknamed striped gophers, they are brownish-colored squirrels with alternating rows of light and dark spots along their sides and backs. The namesake of the Minnesota Gopher athletic teams, they have invaded the entire Great Plains region, from Canada to parts of southern Texas. Strictly daylight marauders, the Thirteen-lined Ground Squirrel prefers tidy areas where grasses are kept mown, such as cemeteries, golf courses, parks, and, of course, yards. They have a taste for meat and will devour grasshoppers, small birds, and even smaller rodents. Their burrow systems are not as complex as those of other ground squirrels, with 15- to 20-foot-long (4.6–6.1 m) tunnels being about average. With a warning call that sounds like the sharp trill of a bird, this is one ground squirrel that is easily identified sight unseen.

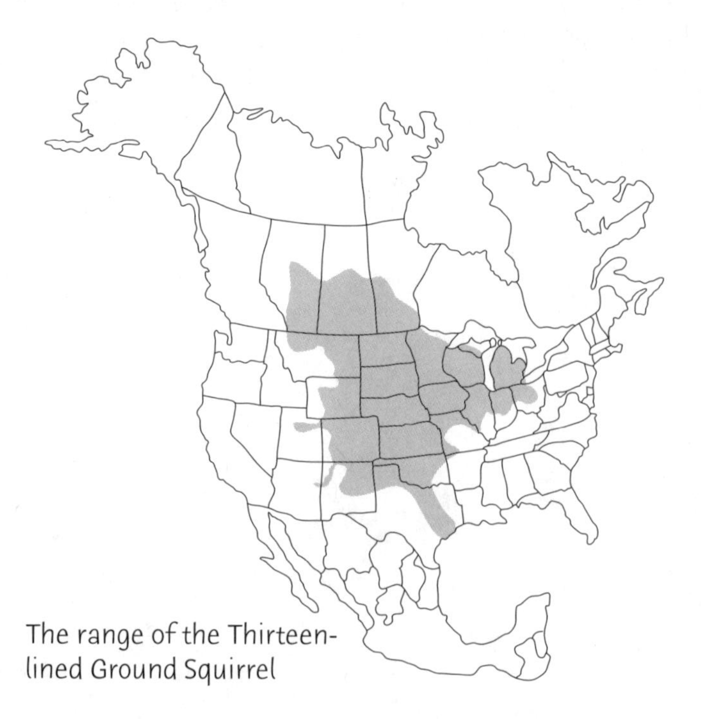

The range of the Thirteen-lined Ground Squirrel

## Thirteen-lined Ground Squirrel Stats

**Average life span**: 2–3 years
**Average size**: Length up to 11 inches (27.9 cm); weight 5.5–8 ounces (156–226.8 g)
**Positive ID**: Brownish to golden-colored with alternating rows of light and dark spots along their sides and backs
**Hibernation**: Varies: Some ground squirrels sleep from mid-October through April; others split the season, estivating during the heat of summer and hibernating through the worst of winter
**Habitat**: Pastures, meadows, suburban yards, parks, gardens
**Range**: South central Canada and central United States
**Signs**: Burrow entrances, sometimes with narrow pathways trampled in the grass
**Territory**: Usually about 50–100 yards (45.7–91.4 m) from the burrow opening
**Number of young born per year**: Usually 1 litter in spring, averaging 3–8 offspring
**Most-active periods**: Daylight hours during spring and early fall
**Diet**: Grass, roots, bulbs, stems, leaves, insects, and, in some cases, eggs and smaller mammals
**Natural enemies**: Birds of prey, foxes, coyotes, badgers, bobcats, snakes, and people
**Positive contributions**: Fertilization and aeration of soil; consumption of harmful insects and weed seeds

The Golden-mantled or Thirteen-lined Ground Squirrel is mistaken often for a chipmunk.

## *Tree Squirrel* (*Sciurus* spp.)

When most people think "squirrels" they think "tree squirrels," those furry furies that take to the trees and cuss us out anytime we wander too near. Like rats, mice, pigeons, and some deer, squirrels are among those adaptable wonders of the animal kingdom that have thumbed their twitchy little noses at us and moved right in. They make themselves at home both in the farthest reaches of the wild and in our attics. They are equally happy foraging for nuts in the deep autumn woods or fidgeting for sunflower seeds at the bird feeder. Blessed with sweet faces, perky ears, expressive eyes, and cute, fluffy tails, these little varmints get away with murder.

But more than just a fluffy tail sets these arboreal acrobats apart from the rest of squirreldom. The tree squirrels, no doubt, fancy themselves the elite of the group. They waste no time in hibernation; most are active all year long, except during torrential downpours. They are specially adapted to a life among the tree limbs, equipped with lithe, muscular limbs of their own, incredible balance and strength, and grappling-hook claws for climbing. But there's more. There's one thing that sets them apart from "lesser" squirrels and makes them such a menace to humankind — a larger brain. Tree squirrel skulls are measurably bigger between the eye sockets, and deeper throughout the braincase, than the skulls of ground squirrels or chipmunks of similar size. With a total cranial capacity about equal to the size of a walnut, tree squirrels are the brains of the outfit.

This might not come as much of a surprise to anyone who has spent time watching their activities. On the other hand, anyone who has laughed at the clown-

### Squirrelly Facts

Most tree squirrels maintain several nests at one time. Once a nest becomes too worn or infested with fleas, the squirrel will leave it and build a new one.

like, seemingly ridiculous antics of a tree squirrel might find this bit of information hard to swallow. In fact, at least two television programs, *Daylight Robbery* and *Daylight Robbery 2,* produced by the BBC in Great Britain, have expounded upon the intelligence, persistence, and creative problem-solving abilities of tree squirrels.

There are several species of North American tree squirrels:

- The Gray Squirrel (*Sciurus* spp.)
- The Eastern Fox Squirrel *(S. niger)*
- The Tassel-eared Squirrel (*S. aberti* and *S. a. kaibabensis*)
- The Red Squirrel *(Tamiasciurus hudsonicus)*
- The Douglas Squirrel *(T. douglasii)*

Another group, the flying squirrels, though obviously tree-dwelling squirrels, are considered separately in the next section.

Just to keep us confused, tree squirrels go by a variety of aliases, some of which cover more than one species of squirrel or vary by geographic region. Western Gray Squirrels are also referred to as California Gray Squirrels; pine squirrel or chickaree applies to both Red Squirrels and Douglas Squirrels, although some folks refer to them as cat squirrels.

**Squirrelly Facts**

Tree squirrels have sweat glands between their foot pads and toes. During periods of hot weather, they leave wet tracks when they walk across a dry surface.

Perhaps even more confounding is the expert camouflage these squirrels have mastered. Gray squirrels can be pure white or jet black, Red Squirrels are usually brown (but black individuals also exist), and fox squirrels come in three different color schemes, with a few in South Carolina sporting white ears. Confused yet?

## Gray Squirrel *(Sciurus carolinensis)*

The relatively large gray squirrel is quite prolific and adaptable, inhabiting nearly two-thirds of the United States. And since these silvery squirrels don't see fit to hibernate in winter, your poor bird feeder is never safe!

### *Eastern Gray Squirrel*

Probably the most common squirrels in the United States, Eastern Gray Squirrels are from 14 to 20 inches (35.6–50.8 cm) in length and weigh an average of 1½ pounds (.68 kg). They are usually gray, with buff-colored fur on their undersides and silver-tipped hair on their bushy tails. Eastern Gray Squirrels can be found in virtually all of the eastern United States, and as far south as Texas, as far west as Oklahoma and Nebraska, and as far north as mid–North Dakota. Having been transplanted into other regions, they can also be spotted in

Gray squirrel families may spend 8 to 30 years perfecting a dwelling in a tree cavity.

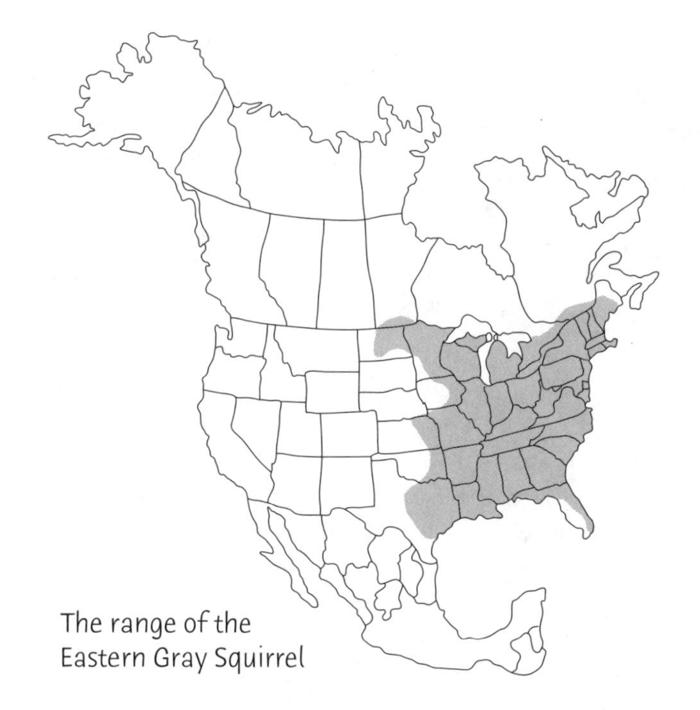

The range of the Eastern Gray Squirrel

some western states, such as parts of Oregon and California. Colonies of pure white squirrels exist in Illinois, New Jersey, and South Carolina. There is also a black form that is most common in the northernmost part of the gray squirrel's range.

Though gray squirrels prefer hardwood or mixed forests that have a good supply of oak and hickory trees, they have come to dominate just about any rural, suburban, or urban setting in which some acceptable food source is available. And they really aren't that picky. They will devour buds, bark, tulip tree blooms, apples, fungi, and many kinds of seeds. Curiously, just like small children, gray squirrels tend to eat one food at a time, picking at the next item only when their first choice is all gone. An individual squirrel typically eats about 2 pounds of food per week, or roughly 100 pounds (45.4 kg) per year.

Squirrels depend on trees for shelter and sustenance, and they are pretty savvy about their real estate preferences. They seek out mature white oaks, beeches, elms, and red

## Squirrelly Facts

North American fossil records verify the presence of gray squirrels as early as 50 million years ago.

maples in search of natural cavities or abandoned woodpecker nests to call home. The process of perfecting a cavity den takes from 8 to 30 years, so it is a matter of generations, not individual squirrels, that create their woodland homes. If the squirrel cannot find a tree cavity at least a foot deep with an opening at least 3 inches around, it may settle for the next-best thing, an intricately woven nest of leaves. Though often obscured by summer greenery, these leafy beds, called *dreys,* are easily spotted in the bare limbs of winter. Less elaborate dreys are also constructed in summer, and the squirrels apparently rotate the use of them, eluding both predators and the buildup of parasites. A sloppily built nest might be the work of a young squirrel or a hurried project meant as temporary shelter.

## Feast or Famine

Squirrel migrations have been documented in many states. The fall of 1967, for instance, saw a high yield of native nuts in Tennessee, Georgia, and North Carolina and, as a consequence, a high birth rate among the resident squirrels. However, the following spring a late frost curtailed nut production, and by October thousands of squirrels were on the move in search of food.

Fall is a busy time for Eastern Gray Squirrels as they hastily collect nuts to store away for winter. Each nut is painstakingly buried, one at a time, just beneath the soil's surface. The squirrel tamps the soil firmly into place with its forepaws to cover the hiding place.

Unlike their ground-bound relatives, Eastern Gray Squirrels and other tree squirrels do not hibernate through the winter. In fact, the very moment that you're refilling your bird feeder for the umpteenth time in late winter is prime time for gray squirrel activity. This is their mating season, and they can put on a spectacular show as they dash through the bare tree limbs, males in hot pursuit of potential mates and rival suitors, females playing hard to get. Female gray squirrels are particular about their consorts, rejecting many a would-be mate. About 45 days after mating, a litter of 2- to 3-inch-long (5.1–7.6 cm) babies is born. The newborn squirrels are blind, hairless, and toothless and typically weigh no more than 1 ounce (28.3 g). Squirrels are also active during the winter months, digging up all those acorns, hickory nuts, and walnuts they so carefully cached the previous autumn. Nuts are recovered, one at a time, via the gray squirrel's exquisite sense of smell. Roughly 85 percent of the stores are recovered. (Still, you can see how the lost 15 percent lends a hand in reforestation.)

Another interesting phenomenon among Eastern Grays is their reaction to overpopulation or insufficient food supplies. Naturalists have observed huge numbers of squirrels on the move en masse, presumably to better feeding grounds or a less crowded area. In the days when their range was heavily forested, these migrations took place through the treetops; the squirrels could cover great distances by leaping from limb to limb. Nowadays, when squirrels start feeling overcrowded, they tend to migrate over land, fording rivers and braving highways.

Perhaps begrudgingly, I must admit that Eastern Gray Squirrels have contributed more than their fair share to the environment. As I mentioned earlier, any one squirrel can

## A Squirrel Never Forgets? Oh, Sure It Does!

It was once believed that the Eastern Gray Squirrel possessed an incredibly accurate memory because of its ability to relocate nuts stored underground months earlier. Researchers who watch and count such things estimate that the animals recover 85 percent of the nuts they bury. However, when the researchers themselves once buried a few nuts, the squirrels were just as apt to recover those as the ones they had buried personally. The little cheaters were simply sniffing out any nuts they could find.

## Eastern Gray Squirrel Stats

**Average life span**: 7–8 years in the wild, up to 20 in captivity
**Average size**: Length 18 inches (45.7 cm); weight 1.5 pounds (.68 kg)
**Positive ID**: Gray to grayish brown, with lighter to white undersides; bushy, silvery-tipped tails
**Hibernation**: None
**Habitat**: Hardwood or mixed forests, especially with hickory and oak trees; also parks, playgrounds, cemeteries, golf courses, yards, college campuses
**Range**: Eastern half of the United States
**Signs**: Nutshells littering the forest floor, gnawed tree trunks, and leaf nests; in winter, holes in earth or snow where nuts have been excavated
**Territory**: Nonterritorial; home ranges vary in size and overlap
**Number of young born per year**: Usually 2 litters of 4 babies per year, one in spring and one in summer
**Most-active periods**: 2 hours after sunrise and 2–5 hours prior to sundown
**Diet**: 97 plant and up to 14 animal items, including nuts, seeds, apples, wild cherries, fungi, buds, bark, bones, antler marrow, bird eggs, nestlings, bird food, and frogs
**Natural enemies**: Parasites, hawks, owls, weasels, coyotes, foxes, and bobcats
**Positive contributions**: Plant trees by burying nuts; add to the diversity of wildlife

be credited with inadvertently planting dozens of trees, so consider the impact of the millions of them currently at work. They also keep the other woodland, park, golf course, and yard creatures apprised of all goings-on, diligently standing guard duty to the benefit of all. When it comes to their contributions to their human cohabitators, at the very least they can be appreciated for comic relief, intellectual challenge, and pure awe as they romp effortlessly along overhead branches.

***Western Gray Squirrel***

Western Gray Squirrels tend to be larger and more distinctly gray in color than their eastern counterparts. They also have white rather than buff-colored bellies. With a fairly small territory, the squirrels are only found in central Washington State and western parts of Oregon and California. Though superb climbers and acrobats, Western Gray Squirrels seem to prefer foraging on the ground. And even though they are ever on guard and quick as greased lightning, their ground feeding puts them at greater risk of predation by coyotes, foxes, bobcats, hawks, and owls.

The squirrels build nests in the trees from 20 feet high (6.1 m) on up, using shredded bark and twigs in addition to leaves. For comfort the squirrels collect moss, feathers, and

The range of the
Western Gray Squirrel

## Western Gray Squirrel Stats

**Average life span:** 7–8 years in the wild, up to 20 in captivity
**Average size:** Length 20 to 24 inches (50.8–61 cm); weight from 14 ounces to more than 2 pounds (.4–.91 kg)
**Positive ID:** Gray to grayish brown, with lighter to white undersides; bushy, silvery-tipped tails
**Hibernation:** None
**Habitat:** Hardwood or mixed forests, especially with hickory and oak trees; also parks, playgrounds, cemeteries, golf courses, yards, college campuses
**Range:** Central Washington, western Oregon, California
**Signs:** Nutshells littering the forest floor, gnawed tree trunks, and leaf nests; in winter, holes in earth or snow where nuts have been excavated
**Territory:** Nonterritorial; home ranges vary in size and overlap
**Number of young born per year:** 1 litter of 3–5 young, in March to June
**Most-active periods:** 2 hours after sunrise and 2–5 hours prior to sundown
**Diet:** 97 plant and up to 14 animal items, including nuts, seeds, apples, wild cherries, fungi, buds, bark, bones, antler marrow, bird eggs, nestlings, bird food, and frogs
**Natural enemies:** Parasites, hawks, owls, weasels, coyotes, foxes, and bobcats
**Positive contributions:** Plant trees by burying nuts; add to the diversity of wildlife

bits of fur or fluff to line the nests. Another common thread with their eastern cousins is their taste for acorns; this often puts them at odds with the California woodpecker, which shares the southern part of their range. The woodpeckers carefully store acorns in the holes of dead trees. When light-fingered squirrels attempt to snatch the birds' bounty, an air-borne battle often ensues.

### *Arizona Gray Squirrel*

Nestled into remote valleys and canyons of Arizona, the Arizona Gray Squirrel is medium sized and plain gray in color with a white tummy and a long, white-fringed tail. It shares its small range with the larger and flashier tassel-eared squirrel, compared to which it looks rather dull. The Arizona Gray Squirrel eats pinecones, acorns, nuts, seeds, and berries. It takes advantage of the deciduous trees in its range, building leafy nests among the cottonwoods, sycamores, and walnut trees.

The range of the Arizona Gray Squirrel

## Arizona Gray Squirrel Stats

**Average life span**: 7–8 years in the wild, up to 20 in captivity
**Average size**: Length 20 inches (50.8 cm); weight 1.25–1.5 pounds (.57–.68 kg)
**Positive ID**: Gray body with white undersides; long, white-fringed tail
**Hibernation**: None
**Habitat**: Hardwood or mixed forests, especially with hickory and oak. Parks, playgrounds, cemeteries, golf courses, yards, college campuses
**Range**: Arizona, New Mexico, and Mexico
**Signs**: Nutshells littering the forest floor, gnawed tree trunks, and leaf nests; in winter, holes in earth or snow where nuts have been excavated
**Territory**: Non territorial; home ranges vary in size and overlap
**Number of young born per year**: 1 litter of 3–4 young in early summer
**Most active periods**: 2 hours after sunrise and 2 hours prior to sundown
**Diet**: Pinecones, acorns, nuts, seeds, and berries
**Natural enemies**: Parasites, hawks, owls, weasels, coyotes, foxes, and bobcats
**Positive contributions**: Plant trees by burying nuts; add to the diversity of wildlife

### Eastern Fox Squirrel *(Sciurus niger)*

Named for its exquisitely plumed tail, the Eastern Fox Squirrel roams along the eastern coast of the United States, though it's rarely found in New England. In fact, fox squirrel populations everywhere have been greatly reduced in recent years.

Fox squirrels have a unique and important relationship with the forest ecology, especially in the southeastern portions of their range. Preferring open areas among the trees, fox squirrels are often found in or near forests that are managed as timber or woodlot land, or in areas that have been

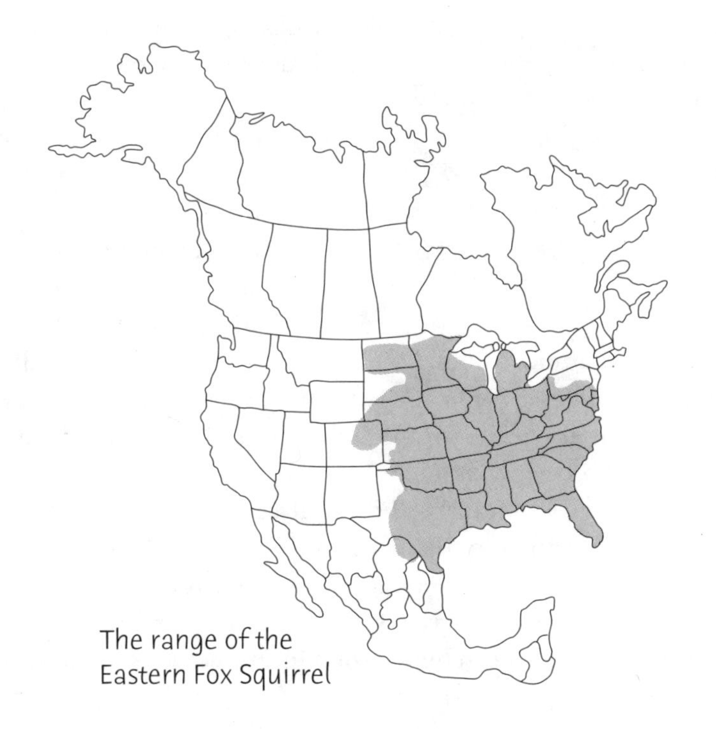

The range of the
Eastern Fox Squirrel

burned. There, they routinely dine upon subterranean fungi. These fungi are loaded with spores that are not digestible, and the fox squirrels distribute them far and wide as they go about their business on the forest floor. The spores, in turn, spread the fungi, which aid in the germination and growth of trees. In a nutshell, the fungi help trees, which help squirrels, which help fungi — a neat circle of cooperation. Unfortunately, the modern-day practice of clear-cutting has removed a vital part of the circle and had a devastating impact on the squirrels.

Fox squirrels are easily identified despite the fact that they are similar to, and coexist in the same general region as, Eastern Gray Squirrels. For starters, they are decidedly larger than gray squirrels. If gray squirrels are the brains of the squirrel community, fox squirrels are the brawn. They are, in fact, the largest of the tree squirrels, often weighing in at nearly 2½ pounds (1.1 kg). They also have thick,

## Squirrelly Facts

The average squirrel has a brain the size of a walnut.

blocky heads, in contrast to the rounded cranium of the gray squirrel.

Fox squirrels differ in color from gray squirrels. In the Northeast, they are gray with a yellowish tummy; toward the western part of their range, their tummy hair is a rusty shade; and in the South, fox squirrels are generally black (melanistic), with white on their faces and tails. They tend to run along the ground when threatened rather than bolting up a tree, which can be unfortunate, since they are slower than their cousins. Fox squirrels are more active in the middle of the day than are gray squirrels, though they have been known to nap on hot summer days.

Of course, fox squirrels are similar to their close kin in many ways. They both tend to den in tree cavities, build leaf nests (each squirrel maintaining several), and spend most of their waking hours in the relentless search for nuts and other forage, including maple seeds, corn, buds, berries, fungi, and green pinecones.

Fox squirrels unknowingly help to replant sparse woodlands by spreading fungal spores that aid the germination of trees.

## Eastern Fox Squirrel Stats

**Average lifespan**: 6 years in the wild, up to 10 in captivity
**Average size**: Length 18–28 inches (45.7–71.1 cm); weight 2 pounds (.91 kg)
**Positive ID**: The biggest of the tree squirrels, with 3 common color patterns: gray above, yellowish beneath; gray above, rust beneath; and (in the South) black, often with a white facial stripe and white-tipped tail; also a blockier-shaped head than gray squirrels
**Hibernation**: None
**Habitat**: Hardwood or mixed forests; live oak, cypress and mangrove swamps; prefers open areas between well-spaced trees
**Range**: Eastern half of the United States; some have been introduced into Washington, Oregon, and California
**Signs**: Nutshells in piles beneath trees or stumps, gnawed tree trunks, large leaf nests; in winter, holes in earth or snow where nuts have been excavated
**Territory**: Nonterritorial; home ranges vary in size and overlap
**Number of young born per year**: 1–2 litters per year, of 2–4 young, in April through September
**Most-active periods**: Late morning through early afternoon, year-round
**Diet**: General feeders, depending on their environment; nuts, flowers, buds, fruits, seeds, pinecone seeds, bones, eggs, frogs
**Natural enemies**: Parasites, hawks, owls, weasels, coyotes, foxes, and bobcats
**Positive contributions**: Plant trees by burying nuts; distribute fungal spores, which aid in tree germination and growth; add to the diversity of wildlife

## Tassel-eared Squirrel *(Sciurus spp.)*

Even though they are found only in a few odd pockets of the Southwest, tassel-eared Abert's Squirrels *(Sciurus aberti)* and Kaibab Squirrels *(S. a. kaibabensis),* the fashion plates of the entire squirrel clan, are worth getting to know.

What sets these squirrels apart from their common cousins are the fancy fringes of their ear tufts. Don't laugh; humans flaunt sillier things. Interestingly enough, this rare feature among North American squirrels has been all the rage among many species of northern European squirrels for aeons.

Isolated long ago during the formation of the Grand Canyon, these stylish squirrels evolved on their own. Over time, they shunned the gray garb of the other squirrels in their area in favor of more elegant attire. Abert's Squirrels sport a blackish gray topcoat, often reddish at the back, with a nifty black border along the sides and a pristine white front. Their lushly plumed tails flare with long black and gray hairs along the top and bright white beneath and along the edges. Others boast basic black all over. Very chic. But it is those fanciful tassels on their ears, just a shade or two darker than

Tassel-eared squirrels lose their characteristic tufts each spring and regrow them each fall.

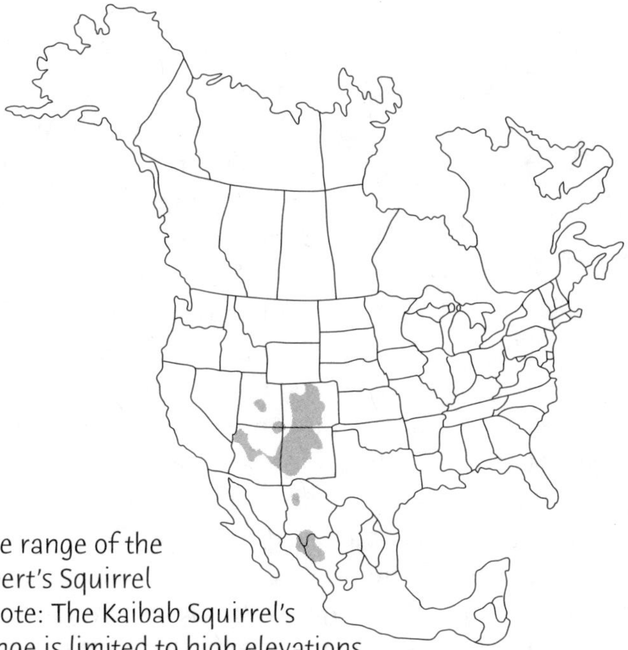

The range of the
Abert's Squirrel
(Note: The Kaibab Squirrel's
range is limited to high elevations
in northwestern Arizona.)

their coat and sprouting straight up ¾ inch (1.9 cm) from the tips of the ears, that have given them their claim to squirrel-fashion fame. They doff the fringes in spring and summer, but replace them anew each fall. They are large squirrels, from 18 to 24 inches (45.7–61 cm) in length and tipping the scales at nearly 2 pounds (.91 kg).

Kaibab Squirrels opted for opposites. While the Abert's Squirrel's tail is dark above and white beneath, the Kaibab wears its white above and black underneath. Its dark gray coat extends all the way around, making for quite a striking contrast with its brightly plumed tail. And again, there are those aristocratic, dyed-to-match ear tufts.

Abert's Squirrels are native to areas from central Colorado through northern and central parts of New Mexico, Arizona, the southeastern edge of Utah, and bits of Mexico. Kaibab Squirrels prefer the high elevation

## Squirrelly Facts

Squirrels can consume a variety of poisonous plants and mushrooms without suffering any ill effects.

(6,000 to 9,000 feet [1.8–2.7 km]) of the North Rim of the Grand Canyon in the far northeastern corner of Arizona. Both rely on a habitat of pine forests, junipers, and piñons. Because their range is so limited, their diets have evolved to match their sparse domain. Primarily pine seed eaters, they also indulge in the bark and buds of the plentiful Ponderosa pines of their area. Piñon nuts and some flowers are also favored fare. In addition, Abert's Squirrels feast on mistletoe and other forms of vegetation.

Like most squirrels, both bury nuts for future use, but abstain from bringing them into their unusual nests. Unlike all other North American squirrels, however, they do not rely on stored food for winter. Instead, they subsist on the inner bark of pine trees. Nest building is dictated by the materials at hand. Natural tree cavities are scarce in their range, so such dens are rare. Leafy trees are few and far between, but pine twigs and witch's broom (pine twigs entangled with mistletoe) are plentiful. The squirrels' peculiar-looking, bulky nests built from these goods can be spotted high in the branches of pine or juniper trees. The nests are about a foot in diameter and, though the outer view prickles with pine twigs, the inside is soft with mosses, lichen, and other padding.

Busy by day, as are most other squirrels, tassel-eared squirrels are active year-round. They are not territorial; each squirrel maintains a large range of up to 60 acres (.24 square km). Tassel-eared squirrels mate only once per year, usually between the months of February and June, resulting in one to five babies. Those that mate earlier in the year produce fewer offspring. Although their life span is of an unknown length, it is probably similar to that of the gray and fox types: seven or eight years in the wild. It *is* known that they stay sexually active throughout their lives. Must be the cool outfits.

## Tassel-eared Squirrel Stats

**Average life span:** 7 or 8 years (estimated)
**Average size:** Length 18 to 24 inches (45.7–61 cm); weight 2 pounds (.91 kg)
**Positive ID:** Varies according to species
**Hibernation:** None
**Habitat:** Pine forests
**Range:** Varies according to species
**Signs:** Nests
**Territory:** Nonterritorial; home ranges vary in size (up to 60 acres (.24 square km)) and overlap
**Number of young born per year:** 1 litter per year, of 1–5 young, in April through September
**Most-active periods:** Late morning through early afternoon, year-round
**Diet:** Specialist feeders: pine seeds, bark, buds, some mushrooms and flowers, nuts where available; Abert's squirrels eat mistletoe as well
**Natural enemies:** Hawks, most notably goshawks and red-tailed; coyotes, foxes, bobcats, and cougars
**Positive contributions:** Add to the diversity of wildlife and look good doing it

### Red Squirrel (*Tamiasciurus* spp.) and Douglas Squirrel *(Tamiasciurus douglasii)*

Red Squirrels *(Tamiasciurus hudsonicus)* and their northwestern cousins the Douglas Squirrels *(T. douglasii)* are the real movers and shakers in the world of squirrels. The genus name *Tamiasciurus* adds the designation "steward" to the general squirrel title of "shade tail." And these shady-tailed stewards, or sentries, of the woodland take their guard duty quite seriously. The noisiest and most active squirrels in the woods, they are pint-sized little characters that pack a giant verbal repertoire and are constantly on the alert.

Also known as pine squirrels, chickarees, barking squirrels, red bobbers, chatterboxes, and *adjidaumo* (a Native American name meaning "tail in the air"), these squirrelly

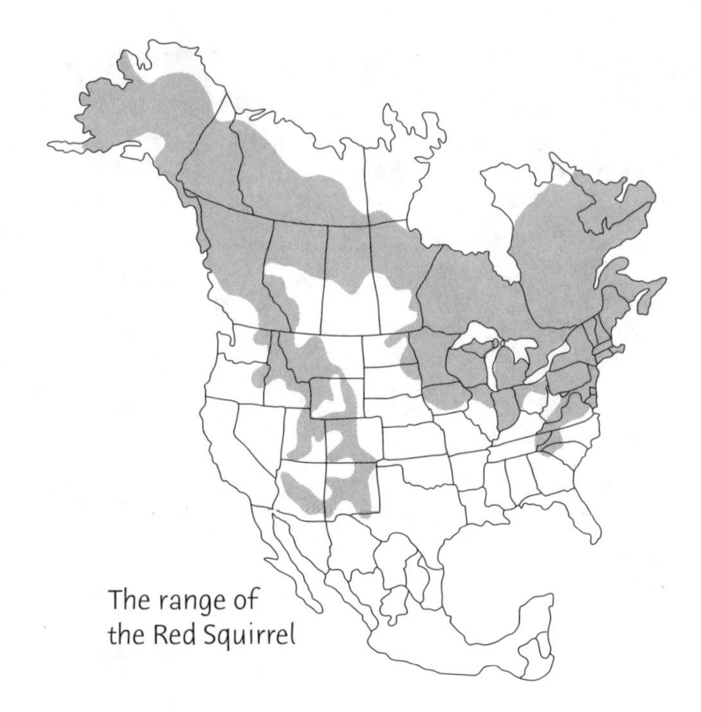

The range of
the Red Squirrel

## A Squirrel by Any Other Name

"The Red Squirrel is a veritable Puck-o'-the-Pines – an embodiment of merriment, bird-like activity, and saucy roguery . . . as boisterous as he is vigorous in work and play."

– Ernest Thompson Seton

cousins are instantly recognizable. They are only about a foot in total length and weigh a scant 5 to 8 ounces (141–226 g) at maturity. Their reddish coats lend them their common name, though the true red squirrel has a more definite reddish tint to its fur, and a more contrasting whitish belly, than the browner Douglas Squirrel. In the northern reaches of their range, they may grow tufts of hair on their eyes in the wintertime. Another identifying feature, their distinctive calls and trills, can be heard almost continuously and for impressive distances, the Douglas Squirrel being the more vocal of the two.

Found throughout much the northeastern and northwestern United States, as well as most of Canada, Red Squirrels are ubiquitous within their range. Almost any type of forest will host a few Red Squirrels, though Douglas Squirrels prefer stands of conifers. Outbuildings, woodsheds, woodpiles, and debris piles entice them to set up housekeeping. Unlike other tree squirrels, they will make use of underground tunnels, as well as master the treetops.

When the weather is good, the Red Squirrel busily gathers and stores food, some items

## On the Douglas Squirrel

"[T]he mockingbird of squirrels, pouring forth mixed chatter and song . . . barking like a dog, screaming like a hawk, chirping like a blackbird or a sparrow; while in bluff, audacious noisiness he is a very jay. . . . He is, without exception, the wildest animal I ever saw — a fiery, sputtering, little bolt of life."

— John Muir

The range of the Douglas Squirrel

buried individually and others neatly arranged in tree forks or along branches. Red Squirrels also hoard their prizes in *middens,* which are huge, heaping piles of pine- or fir cone scales, nut husks, twigs, and bark. Middens are typically sited beneath a favorite perch or at the base of a stump. Some have been measured at 20 feet (6.1 m) long, 12 feet (3.7 m) wide, and 3 feet (.92 m) deep. Large middens are typically riddled with tunnels and nest cavities.

The Red Squirrel has an uncanny knack for knowing how and when to harvest, store, and eat its food. For instance, with a typical squirrelly sweet tooth, the Red Squirrel knows which trees offer the sweetest sap, and that by gnawing off a bit of bark and a slight hollow beneath, the depression will fill with the sweet liquid. The squirrel also knows that by letting the sap sit until the moisture evaporates, the sticky puddle becomes much sweeter. Another smart tactic that this critter employs is to cut green cones before they ripen and stow them under a layer of humus or pine needles. The cones ripen, seeds intact, ready for the hungry squirrel come a cold winter's day.

Red Squirrels roam over a vast and varied terrain, and so the kinds of food they consume also tend to vary. As a result, the squirrels divvy up their spoils according to two basic categories: soft food that will go bad quickly and must be eaten first, and hard food that will keep. White pine is a staple of their diet; berries, fruits, and wild roses are favorite treats, as are herbs and fungi. The squirrels enthusiastically collect, purposefully dry, and greedily consume various kinds of mushrooms, including poisonous amanitas, with no ill effects. Buds, soft stems, seeds, and the cones of pine, larch,

### The Makings of a Meal

One Red Squirrel can strip and consume the seeds of 12 pinecones in a single setting. The animal can strip a cone in just under two minutes, proceeding from base to tip and discarding the inedible portions with a quick flip of its paw. It will consume an average of 540 seeds per meal.

To satisfy a sweet tooth, Red Squirrels "tap" trees by chewing sap-collecting hollows in the bark.

spruce, willow, aspen, poplar, and many other trees are fair fare. They often stash bits of antler or bone in their middens, and they even have a taste for meat. They will kill small birds, rabbits, and even young gray squirrels. Many insects and invertebrates, such as snails, also find their way to Red Squirrel stores.

It seems that just about anything a Red Squirrel touches is deemed its own. In fact, relocating, licking, or otherwise manipulating any food item legally defines it, in the eyes of the squirrel, anyway, as squirrel property.

Red Squirrels of both types are extremely territorial, jealously guarding their domains and food stores from friends and foes alike. Female Red Squirrels tolerate the company of a mate for only a single day in late winter, and otherwise share their territory only when raising young. A Red or Douglas Squirrel knows every

## The Wanderers

Like gray squirrels, Red Squirrels have been known to stage some spectacular migrations, in some cases running, climbing, or even swimming extreme distances in search of a new home. One case of a tagged and tracked young male found the squirrel more than 70 miles (112.9 km) away from his birthplace.

## Red and Douglas Squirrel Stats

**Average life span**: Unknown
**Average size**: Length 10–14 inches (25.4–35.6 cm); weight 5–8 ounces (141–226 g)
**Positive ID**: Varies according to species
**Hibernation**: None
**Habitat**: Mixed and conifer forests
**Range**: Red Squirrels are in Alaska and Canada, south through the Rocky Mountains, and in the northeastern quarter of the United States; Douglas Squirrels roam southern British Columbia, western Washington and Oregon, and northern California
**Signs**: Middens (large piles of conifer scales and nut husks with an entry hole at one end)
**Territory**: 2–3 acres (8094–12,140 square km)
**Number of young born per year**: 1–2 litters per year (1 in spring, another in late summer) of 3–7 young
**Most-active periods**: Late morning through early afternoon, occasionally after dusk, year-round
**Diet**: General feeders: nuts, flowers, buds, fruits, seeds, pinecone seeds, bones, eggs, small birds, rabbits, gray squirrels, frogs
**Natural enemies**: Parasites, pine martens, fishers, minks, weasels, bobcats, domestic cats, hawks, owls
**Positive contributions**: Plant trees by burying nuts and conifer seeds; eat tree pests

inch, up, down, and sideways, of its 2- to 3-acre (8094–12,140 square km) territory. The squirrel is forever on the move, checking out every nook, cranny, limb, tree fork, and tunnel in a constant exploration of the available menu. An intimate knowledge of its surroundings means the squirrel is always just a hop, skip, or a jump away from safety should a predator threaten. It is fearless, independent, curious, and acutely aware of its self-designated property rights, as anyone who has dared trespass upon its turf is no doubt acutely aware!

Active all year, Red and Douglas Squirrels are primarily diurnal but have been known to challenge flying squirrels in the dark of night to nut-husking forays. Small, fast, and agile, the Red Squirrel suffers from few predators. However, the pine marten is a prime exception, as are hawks, owls, fishers, minks, weasels, bobcats, and domestic cats. If and when this tiny swimmer takes to the water, it becomes an unfortunate lure to large, hungry fish. It is also often bedeviled by pests and parasites, including fleas, mites, ticks, and botflies, which lay their eggs on the squirrel's fur to hatch and then burrow under the skin. Mice routinely rob their stores, and in the case of the Douglas Squirrel, whose caches of conifer seeds are legendary, humans are often guilty of pilfering their reserves.

## *Flying Squirrel* (*Glaucomys* spp.)

The silent gliders of the night skies, flying squirrels sail noiselessly from tree to tree, their movement so discreet that few know of their presence. Possessing oversized, ringed eyes — their version of night-vision goggles — they are the only squirrels that work a full night shift. By morning's light, they have again taken to their dens in hiding.

The genus is named for the general appearance and size of these tiny night sprites, translating into "gray mouse." They are soft gray to grayish brown in color with white or light-colored undersides. Small in stature, they are no more than a foot in length from nose to tail tip, and they often weigh less than 3 ounces (85 g) at maturity. There are two species, the Northern Flying Squirrel *(Glaucomys sabrinus)* and the Southern Flying Squirrel *(G. volans),* the smallest of North American tree squirrels.

Most people don't realize that flying squirrels are in their midst, but only the southwestern quarter of the continent is without them. The Northern Flying Squirrel inhabits areas from the lower two-thirds of Canada and eastern Alaska down through the Great Lakes, New England, Washington State, and Idaho. The Southern Flying squirrel has invaded most of the eastern United States with the exception of

"Flying" squirrels glide with the help of a fold of skin called the *patagium* to reach a lower perch 30 or more yards (27.4 m) away.

Maine, the southern tip of Florida, and far northern Minnesota. In the areas where their ranges overlap, around the Great Lakes and parts of New England, their different habitat preferences still make it possible to tell one variety from the other.

The Northern Flying Squirrel seeks out coniferous or somewhat mixed forests, while its southern cousin prefers stands of deciduous trees. Tall deadwood stumps are favorite nesting places, as are trees pocked with woodpecker holes. Lined with shredded bark, moss, feathers, fur, and other soft material, even nests in outbuildings, barns, attics, and birdhouses make comfy quarters. In a pinch, the squirrels will construct twig-and-leaf nests similar to, though much smaller than, those of the gray squirrel.

Finding flying squirrels in the wild calls for some clever espionage. Northern Flying Squirrels often leave piles of cones at the base of their nest trees, while Southern Flying Squirrels leave hard-to-find clues of their presence in the form of hickory nuts with smooth openings gnawed into the thin end. Normally active only at night, they are most often exposed in broad daylight when their tree homes are cut down. Tapping on the trunk of a likely looking snag may cause the curious occupant to peer through its front door, or even to flee. In the dark, they are often heard and not seen. Though eerily quiet in flight, their high-pitched voices are frequently confused with the shrill calls of night birds. As nestlings, their squeaks sometimes reach into the same ultrasonic range as those of bats.

Contrary to their common name, these squirrels don't actually fly — they fall. Leaping from a high branch, the flying

### Social Animals

Gregarious by nature, flying squirrels differ from many of their cousins in their delight with roommates. They rejoice in group flights, and it is common to find several squirrels holed up together for the night. As many as 50 have been found sharing a single nest cavity, though 4 to 15 is more common.

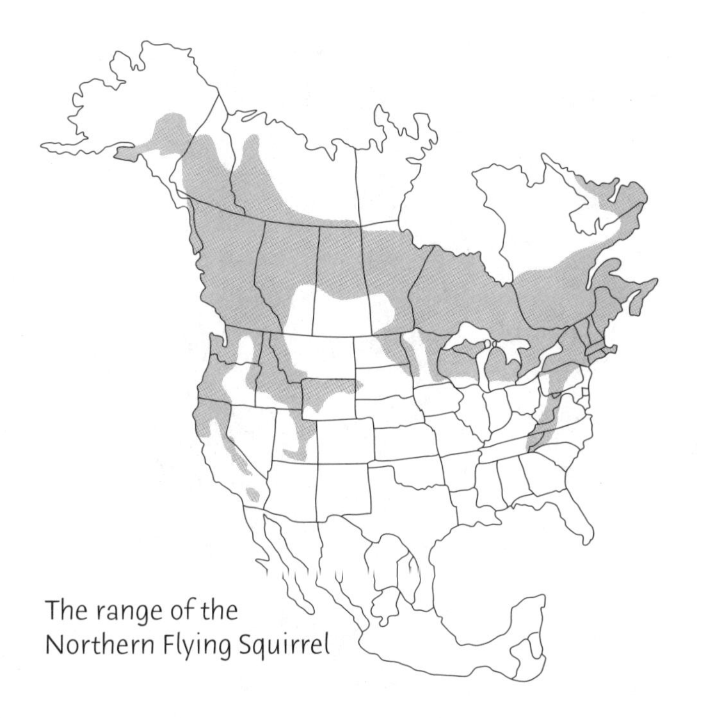

The range of the Northern Flying Squirrel

> "Oh-oh, Bullwinkle. I think we're in trouble!"
>
> — Rocket J. Squirrel, flying squirrel of undetermined species

squirrel spreads its legs wide, stretching from the wrist of the front leg to the ankle of the hind leg a fold of skin, the *patagium,* that acts as both parachute and glider wing. Down, down, down the squirrel sails, twisting to avoid a tree limb here, adjusting its angle of descent there, and suddenly, with an abrupt pull-up, it alights on a lower tree trunk far from its point of departure. Quickly, the squirrel darts to the other side of the tree in a reflexive effort to elude any predator that might have watched its brief flight. An average flight is about 20 to 30 yards (18.3–27.4 m), but leaps of more than 100 yards (91.4 m) are not uncommon. Though they can alter their flight paths somewhat after takeoff, the ratio of descent to horizontal ground covered is about 1:3. This means that for every foot they fall, they cover 3 feet (.92 m) of ground.

The range of the Southern Flying Squirrel

Quick and confident in the air and on tree limbs, flying squirrels are reduced to comical clumsiness on the ground. However, they spend a good deal of their nightly feeding forays foraging on the forest floor, and at least one squirrel has been documented to have romped more than ¼ mile in returning to its nest across a treeless expanse. Knowing every inch of their 4- to 5-acre (16,187–20,234 square m) territory, the squirrels are ever prepared to instantly slip undercover to avoid an enemy.

The Southern Flying Squirrel exploits a variety of foods, including acorns, seeds, and nuts, storing away more than 15,000 per season. It also indulges a sweet tooth, searching out berries or licking at sweet sap. Many are the daunted syrup collectors who, in their routine checking of syrup buckets, find at least one drowned, greedy squirrel having failed in its attempt to satisfy its craving. But oddly, this tiny tyke also has a reputation for being the most carnivorous of all squirrels, actively hunting insects, birds, and eggs.

Northern Flying Squirrels subsist on a diet of lichens and underground fungi, perpetuating the same cycle as the Eastern Fox Squirrel in spreading the spores that in turn propagate the beneficial fungi throughout its territory. They also consume nuts, seeds, and insects, storing away a good portion for winter use.

Southern Flying Squirrels differ from their kin to the north in several other ways. The southern species tends to produce two litters per year, some females experiencing their single day of heat for the year in early spring, others not until midsummer. Northern squirrels, in contrast, raise but one family of two to five babies each spring. Though the slightly larger northern variety is active throughout the year, the southern squirrels tend to enter a form of false hibernation, or torpor, as their body temperatures drop well below freezing (one report states to 22°F [-5.6°C]) for the worst of the winter weather.

## Flying Squirrel Stats

**Average life span**: 3–5 years
**Average size**: Length 10–12 inches (25.4–30.5 cm); weight 1.5–3.5 ounces (42.5–99.2 g)
**Positive ID**: Southern squirrels are tiny, with a silky gray-brown coat, white belly, large fold of skin from front to rear ankles, flattened tail, and big, dark eyes; the northern variety has soft, deep brown fur with a white belly, flattened tail that's dark above but white beneath, and large, luminous eyes
**Hibernation**: Southern squirrel enters state of torpor during extreme cold or periods of insufficient food; northern species does not hibernate
**Habitat**: Southern prefers mixed, deciduous forests; northern likes mixed, coniferous forests
**Range**: Southern is found in eastern United States except extreme southern Florida and Maine; northern in Canada, eastern Alaska, northwestern Oregon, the Great Lakes region, and New England
**Signs**: Trees with woodpecker holes; hickory nuts with smooth holes in thin end; piles of gnawed cones at the base of trees
**Territory**: Very gregarious; individual ranges average 4–5 acres (16,187–20,234 square m)
**Number of young born per year**: Southern produces 2 litters of 2–3 young per year, some breeding in spring, others midsummer. Northern squirrels have 2–5 babies in spring or summer
**Most-active periods**: Night, all year (except very cold spells for Southern Flying Squirrels)
**Diet**: Nuts, seeds, sap; Southern Flying Squirrels also hunt insects, birds, and eggs
**Natural enemies**: Martens, fishers, mink, weasels, domestic cats, owls
**Positive contributions**: Plant trees by burying nuts and conifer seeds; eat tree pests

## ***Chipmunk*** (*Tamias* spp.)

Chip and Dale, the cutesy little chipmunks of cartoon fame, were a sham! Not that clever chipmunks couldn't possibly have come up with some of the exploits that those two perpetuated. Not that they might not even have chattered away at one another, but cooperate? Chipmunks? No way.

Of the 22 species of chipmunks, 21 of which are native to the western half of the United States and one, the Eastern Chipmunk *(Tamias striatus),* to the eastern half of the country, not one of them is polite to the others. They are all members of the same genus, *Tamias,* which means "storer," "steward," or, in a somewhat looser translation, "paranoid little rodent that can't get along with its neighbors." Perhaps in a salute to the varmints' headstrong nature, the Algonquian Indians named them chipmunk, which translates to "headfirst."

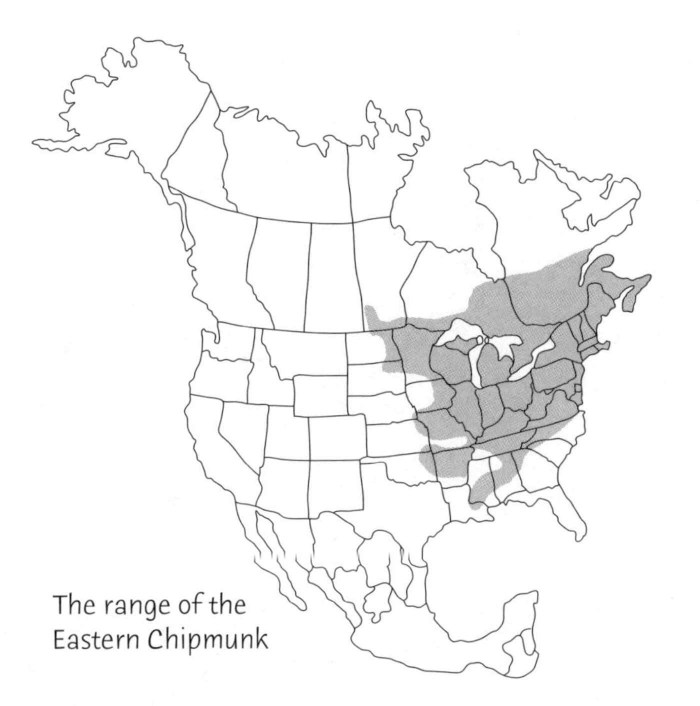
The range of the
Eastern Chipmunk

Chipmunks are wood sprites. They scamper merrily beneath the cover of brushy areas, their striped fur blending with the filtered lights and shadows that caress the forest floor. Those native to drier areas tend to be lighter in color with indistinct striping patterns for camouflage in desert terrain. Natives of the open forest have more clearly defined stripes to belie their presence. One species or another can be found almost anywhere in North America. In fact, the only places that chipmunks have not made home are the tropical parts of Florida, areas with swampy soils, open prairies, and north of the Canadian tree line.

It's not hard to recognize a chipmunk, but in the western United States it can be next to impossible to tell one species from the next. Small in stature, the largest chipmunk barely exceeds a foot in length, and most weigh in at 2 to 3 ounces (56.7–85 g). Chipmunks are most readily identified by their snazzy racing stripes. Even those species that have obscure or occasionally absent stripes, such as the Cliff Chipmunk

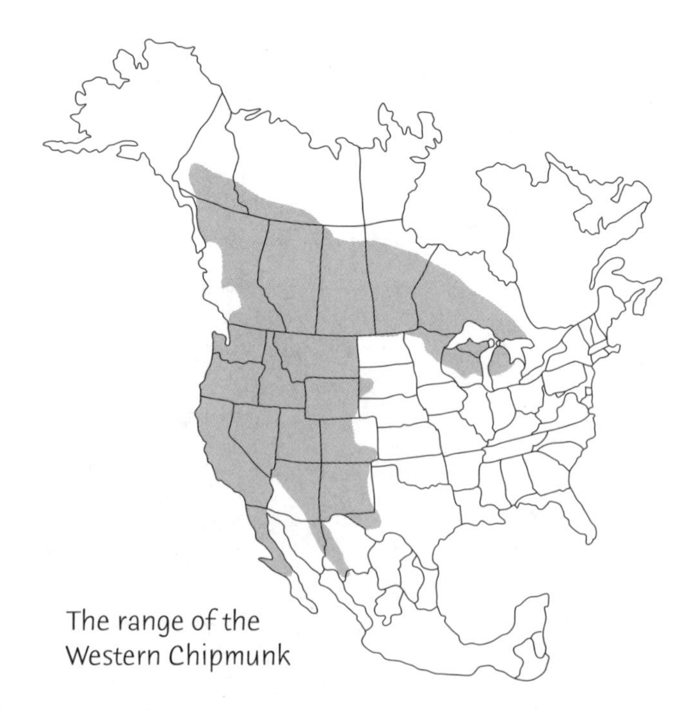
The range of the Western Chipmunk

*(Tamias dorsalis)* or Merriam's Chipmunk *(T. merriami),* tend to have contrasting light and dark stripes around the eyes. All chipmunks have pointy noses, pert, upright ears, and beady little eyes. Western Chipmunks are often brazen little acrobats that will steal your heart with their antics while snatching a snack from close range. Eastern Chipmunks are a bit more reserved. They have clearly marked black and white body stripes, unlike the Thirteen-lined Ground Squirrel, whose stripes mingle with alternating light and dark spots. Eastern Chipmunks are much larger than the only other chipmunk in their range, the Least Chipmunk, which is emblazoned with four distinct side stripes as compared to the Eastern Chipmunk's two. Some chipmunks sport bright, contrasting stripes of rich gold, red, black, and white, while others exhibit more muted shades. Most have bushy tails. Those in cool, moist climates tend to be larger, darker, and more distinctly marked than the rest of the species.

## How the Chipmunk Got His Stripes

One night, a council of forest creatures was called to determine whether daylight or darkness would prevail. The discussion quickly became violent. Chipmunk chimed in in favor of light. Bear sang his praise of darkness. Chipmunk in his characteristic way taunted and teased, demanding the light to return. As the morning dawned, some of the animals were extremely angry with Chipmunk, and having taken all he could stand, Bear ran after Chipmunk and grabbed for him just as he disappeared down a tunnel, chittering and chattering the whole way. But he was not unscathed, and to this day wears the black and white streaks left by the claws of angry Bear.

– Native American Legend

Like the relatives for which they are most often mistaken, such as the Thirteen-lined and Golden-mantled Ground Squirrels, chipmunks are burrowers. Entrances to chipmunk residences are usually well hidden beneath a stump, woodpile, or outbuilding.

## What, No Kitchen Sink?

Chipmunks cram all sorts of stuff into those pouches. The stash of one Yellow-pine Chipmunk *(Tamias amoenus)* of the Pacific Northwest reportedly included more than 67,970 separate items, including 15 types of seeds, corn, and parts of a bumblebee.

During the construction of their homes, chipmunks will even go to the trouble of stuffing their cheek pouches with excavated soil for distribution away from the opening, a bit of landscaping savvy that avoids advertising the opening to would-be predators. Burrows tend to run 20 to 30 feet (6.1–9.2 m) in length, and they often include a master bedroom, several storage areas, and separate escape tunnels. Though most chipmunks spend the majority of their time on or in the ground, several varieties, including the Red-tailed Chipmunk *(Tamias ruficaudus)* of northern Idaho and the Colorado Chipmunk *(T. quadrivittatus),* take to the trees as naturally as any of their squirrelly cousins. And all chipmunks are willing and able to climb in search of acorns, fruits, or other sweet rewards.

Most chipmunks claim a territory of about ½ acre (2023 square m), and they will staunchly defend the immediate area of their dens against intruders. The exception is during courtship, which occurs twice a year, in early spring and

Chipmunks' stripes vary in color and pattern with their native terrain.

again in late summer. Oddly, despite their solitary nature, as many as 15 chipmunks may share an acre of ground if food supplies are sufficient, though two to four per acre (4046 square m) is more the norm.

Flexible, fur-lined cheek pouches are the hallmark of chipmunks. Like other ground-dwelling squirrels, they cram their mouths full of seeds and other bits of food bound for storage. With its cheek pouches full, a chipmunk's head appears double its normal size.

Chipmunk diets consist mainly of seeds, including thistle seeds and pinecone seeds, nuts, and berries, but they relish the occasional treat of mushrooms and bugs, taking cruel delight in pinning down and devouring dragonflies and butterflies. They even enjoy a tasty bit of roadkill from time to time.

## Chipmunk Stats

**Average life span**: 3–4 years in the wild; 12 in captivity
**Average size**: Length 7–12 inches (17.8–30.5 cm); weight 1.5–3.5 ounces (42.5–99.2 g)
**Positive ID**: Varies with species. All have some marking around the eyes and run with their tails carried upright
**Hibernation**: Varies with species
**Habitat**: Woodlands and brushy areas
**Range**: Eastern Chipmunks are in most of the eastern United States and southeastern Canada; western varieties inhabit most of the western United States and the remaining reaches of Canada
**Signs**: Remnants of gnawed acorns or nuts; hidden burrow entrances with no soil scattered around the opening
**Territory**: ¼–½ acre (1011–2023 square m)
**Number of young born per year**: 2 litters per year of 3–7 young, in late spring and fall
**Most-active periods**: Late morning and late afternoon
**Diet**: Nuts, seeds, bulbs, berries, fungi, insects, snails
**Natural enemies**: Weasels, minks, hawks, foxes, bobcats, and domestic cats
**Positive contributions**: Eat weed seeds and insect pests

Chipmunks are day laborers, most active after daybreak and again at dusk. They spend their waking hours compulsively collecting food stores and caching them away. Most are active year-round, though in severe cold some will go into a state of dormancy for two weeks or more at a time. Rather than putting on fat to ward off low temperatures, slumbering chipmunks emerge every couple of weeks or so to plunder their stores, then head back to the boudoir.

Masters of verbal communication, chipmunks have extraordinary hearing as well. They have been known to congregate in song and express themselves in a range of vocalizations. This gives some credibility to the cartoon characters Simon, Alvin, and Theodore Chipmunk, who bickered and quibbled, but loved to chime into song with their high-pitched voices.

"A squirrel leaping from bough to bough, and making the wood but one wide tree for his pleasure, fills the eye not less than a lion, is beautiful, self-sufficing, and stands then and there for nature."

— Ralph Waldo Emerson, "Art"

CHAPTER 4

# Revenge of the Squirrels: The Damage They Do

**Okay. So they're well equipped,** clever, and ever present. Come now. They are, after all, just tiny fluffs of fur. What serious damage can a squirrel really do? Oh, did I forget to mention that when it comes to petty theft, frustration, and property damage, squirrels are also mighty creative?

Several categories of destruction are wrought by these furry fugitives, most of which cost us financially and all of which cost some measure of our peace of mind. Bird enthusiasts probably top the list of those frustrated by greedy squirrels. Gardeners come in a close second, or perhaps tied with those whose chimneys, attics, or ducts have been invaded by the little beasts. Other forms of vandalism are often cause for alarm, and, in some areas, squirrels are known to have an impact on other wildlife. In extreme cases, squirrels have been known to go nuts and commit out-and-out assault.

## *A Bird Feeder's Nemesis*

Bird feeding is one of the most popular hobbies in America today. As a pastime, it leaves baseball, playing golf, jogging, sewing, bicycling, and visiting the Empire State Building in the dust. In 1991 (the latest year for which accurate figures are available), 63 million Americans fed wild birds at home, spending over 18 billion dollars on bird-related activities. There are several bird-feeding magazines, and half a dozen very profitable bird-feeding mail-order companies. Wild Birds Unlimited, one of several national chains of bird-feeding stores, was founded in 1981 and now has more than 240 stores — and it's growing. Bird-feeding books routinely sell hundreds of thousands of copies. Bird feeders even have their own Web site (www.birdfeeding.org).

Americans alone spend billions of dollars each year on bird-feeding supplies, and squirrels reap many of the benefits.

One interesting thing about our penchant for birds is the correlation between bird feeding and our economy. Don't think the squirrels don't know this. Somewhere out there, a lone squirrel sits in its tree tallying all the day's expenditures for bird feeders and seeds, chuckling as yards of adding-machine tape coil beneath it.

As a culture, we are spending more and more on home improvement and home-based activities. We're becoming a nation of homebodies: When the economy goes bad, we scrimp and put our pennies into our homes as a secure investment that uplifts our lives at the most basic level. When the economy picks up, we lavish rewards on ourselves by — what else? — sprucing up our homes. Perhaps a little landscaping, a fancy new birdbath, and hey, while we're at it, how about a few nice bird feeders so we can encourage the local feathered fauna?

And of course there is one more common denominator among those who enjoy the twitter of little beaks at their feeders. Virtually everyone who feeds birds has problems with squirrels. Not surprisingly, many bird-feeding enthusiasts get good and riled at the notion of squirrels pilfering their birdseed. Squirrel-proof products and bird feeders are big business. But why squander a fortune on products when a little cheap understanding will combat the problem much more effectively?

Not all of the bird food that's lost is consumed by the little varmints. Far more annoying than accepting that at least the little buggers are making some use of your sunflower seeds is finding evidence that the beasties merely waste it. A common complaint is that squirrels will empty stores of foods they don't like, apparently merely to inform us of their disapproval. Scattered seeds at the base of the feeder is a clear clue that "Nyah, nyah . . . nyah, nyah, nyah — I can still get to the feeder. You might as well put something worthwhile in there!"

Of course, lack of easy access never did discourage the truly squirrelly at heart. As a last-ditch effort, squirrels that can't actually climb onto a feeder will launch themselves at it, apparently attempting to smack the darn thing hard enough to

**Squirrelly Facts**

After mating with a male squirrel once, a female squirrel will never mate with that same male again.

## Is Nothing Sacred?

Gwen Steege of Massachusetts prized a hand-thrown bird feeder that her son, a potter in California, had made for her. One late-winter afternoon, she happened to spy a gray squirrel doing its darnedest to get to the feeder. "Thirty seconds later," Gwen reports, "crash! My poor feeder was shattered."

dislodge some seeds. The battered burglars then happily scarf up the scattered food and romp back to repeat the whole process. Tough varmints.

Another complaint about aggressive squirrels monopolizing bird feeders is that they scare the birds away. You won't often find birds sharing a feeder with a squirrel, and worse yet, because squirrels are no more inclined to share feeders with *each other,* you might well wind up with a single squirrel at every bird feeder you own. Since squirrels are notorious enemies of birds, often raiding their nests in spring and eating their eggs or even killing fledglings, many birds instinctively avoid squirrel territories. An active, vocal squirrel is reason enough for many birds to shun the area altogether. The rudest of bird-harassing squirrels will add the insult of property damage by nibbling away at the entrance holes to bird houses. Once the hole has been enlarged sufficiently, the furry critter creeps in and gorges on eggs or helpless baby birds.

Should your feeder be visited at night by a flying squirrel, get out the chain saw. Most flying squirrels can comfortably glide a good 50 yards (45.7 m) from a launching point above your feeder and land with pinpoint accuracy. As a result, any nearby tree (or high point) that is 60 feet (18.3 m) or higher poses a threat to your bird feeder. Once a flying squirrel lands on your feeder, it quietly gorges itself on seeds, suet, or peanut butter before making a quick getaway back to its tree. Surely your neighbors down the street will understand if their trees must go.

## It Gets Worse

Having a feeder-dependent population of squirrels around has some unforeseen consequences. Artificial food sources create an unnatural balance in the squirrel:tree ratio. Normally, a habitat will only support as many squirrels as there are sources of food and shelter. If for some reason food becomes available from a source other than trees, however, more squirrels fill in the population gap. But where are these excess squirrels supposed to live? Odds are they'll find a nice cozy spot somewhere very near the food source, perhaps right over your head (attic) or under your nose (basement). And if for any reason the feeder runs dry or is removed, miffed squirrels will turn to whatever is nearby to satisfy their hunger — your garden, flowerpots, dog dish, and so on.

# *Garden Gambits*

Not troubled by tree squirrels? Then perhaps you're battling ground squirrels, which are found in virtually all regions of the United States and will stop at nothing to satisfy their digestive demands and tunneling tendencies. Ground squirrels aren't picky; they've been known to sample everything from garlic roots to bird eggs. And woodchucks, basically just really big ground squirrels, are renowned for their destructive adventures in gardens.

Among the mildest of gardening concerns spurred by squirrel activity is the mysterious appearance of unidentified, or unwanted, plants. Whether it's frustrated squirrels scattering seeds from feeders or industrious individuals purposefully planting, squirrels often seem to add their own little touches to the landscape. Though some birdseed is supposedly pretreated so it will not sprout, many are the bird-feeding gardeners who have learned to expect the emergence of rogue sunflowers.

Digging is another problem. Gardeners who may be content to contend with a few odd plants popping up here and there often fail to see the humor in holes. Tree squirrels burying and retrieving their stores can leave small, ragged depressions throughout your yard. They also seem to have a weird attraction to digging up potted plants. And then there are the ground squirrels. Unlike gophers or moles, which leave mounds and earth trails, ground squirrel holes may be hard to detect. The clever critters hide them at the base of bushes or weeds, leaving little evidence of their existence. They also riddle the subsoil with tunnels. Imagine the headaches they can cause in golf courses!

Woodchucks, on the other hand, construct immense mounds that have a hole in the center, allowing them access to their burrow systems. The mounds are unsightly and pose a threat to livestock and machinery. The tunnels that extend from the mounds are usually 2 to 4 feet (.6–1.2 m) beneath the soil's surface and can go on for more than 20 feet (6.1 m).

## It Gets Worse

By far the most upsetting squirrel gambits in the garden are theft and vandalism. Since most squirrels have a sweet tooth, they will climb over or dig under most obstacles to

Ground squirrels, such as the thirteen-lined variety, damage the garden from above and below.

find a snack to their liking. Tree squirrels are notorious for pilfering pecans, absconding with apples, borrowing berries, and, in short, helping themselves to any small, sweet treat in the garden. They also have a taste for tomatoes, corn, flower bulbs, and freshly sprouted seeds. They damage ornamentals and trees by chewing tender buds, nipping twigs, or stripping bark, and will nibble holes into some trees in an effort to get to the sweet, gummy sap.

Ground squirrels and chipmunks will help themselves to a range of garden goodies, from tender spring greens to late-summer strawberries. They seem especially fond of sprouted seeds from the cucurbit family, such as cucumbers, squash, and melons, as well as newly planted or maturing corn. They also enjoy sampling random vegetables and flowers. Ground squirrels are infamous for their ruination of grain crops and, like woodchucks and prairie dogs, will gnaw grasses down to a stubble. Woodchucks go for legumes, such as field alfalfa, peas, and beans, and are especially attracted to soybeans. Like other squirrels, woodchucks will travel the farthest out of their way for sweet fruits. Ripe melons are a woodchuck beacon.

### Squirrels of the Corn

A keen observer can ascertain the variety of squirrel that is raiding a corn crop merely by inspecting the damage. Fox squirrels like to tear off whole cobs and haul them to a comfortable feeding spot, the base of which is likely to be littered with the mess from their meal. Gray squirrels nibble only the germ end of the kernels from the cob. There may also be a haphazardly constructed leaf nest in a nearby tree.

Ground squirrels and chipmunks assault the entire stalk, climbing to the top to reach the ears. They dislodge the kernels one at a time, pop them into their cheek pouches, and squirrel them away. Woodchucks devour sweet corn by pulling down the entire stalk.

## *The Furry Chimney Sweep*

If you've ever had a squirrel, or worse yet a family of squirrels, trapped in your chimney, you know they're not exactly Santa's little helpers. Most often, squirrels haven't descended your chimney by intention. They can fall in by accident, then must face the daunting problem of getting back out. This brings up two important questions to a squirrel: Where does the chimney lead? and How do I get out of here?

Because the insides of chimneys are generally smooth, the surface may not provide a good-enough toehold for a panicked and possibly injured squirrel to climb back out. If the chimney leads to an oil or gas burner, the squirrel will wind up wherever the burner is kept, usually in a basement. If the flue leads to a fireplace, and the poor rodent makes it this far, it may escape through a damper or directly out of the fireplace to run wildly through your house.

If you have a squirrel in your chimney, chances are it's an accident, not an invasion. Given a rope to grab onto, the critter might retreat.

## Making Themselves at Home

> **Squirrelly Facts**
>
> Some squirrels are very territorial, while others are not. Red Squirrels, prairie dogs, and chipmunks are particularly protective of their territories.

Once in your house, most squirrels find the dim, cozy attic quarters quite appealing, after some minor alterations, of course. Those that do not gain accidental access via the chimney often enter houses as the approaching cold weather in winter drives them to seek new shelter. Juvenile males, exiled from their home range in summer, often find refuge in the comfort and convenience of an attic. After dropping onto the roof via overhanging tree limbs, squirrels enter the attic through vents, gaps in construction, or rotted fascia boards. Sometimes small gaps give them the start they need to chew an entry hole, which allows cold air and humidity into the attic along with the squirrel.

Once in residence, squirrels begin to make themselves at home. They chew up insulation or wood to make nice cozy nests, gnaw through wiring just for the heck of it, and leave little calling cards throughout the area. Needless to say, chewed wires are an electrical-fire hazard.

Finding out that squirrels have invaded your attic may come as more of a surprise than learning you have bats in your belfry. You generally won't see them unless you go looking for them intentionally. You probably won't smell them until they die. But you *will* hear them scurrying about. And since they're active all year long, don't expect a break in the festivities. Most tree squirrels are active in the daylight, their noisy periods being sometime between dawn and dusk. Flying squirrels, however, are nocturnal and may wake you throughout the night. A thorough investigation for evidence might turn up scattered fecal pellets left by gray squirrels or a designated toilet area if the intruders are flying squirrels.

Confronting the culprits is much easier if they are gray or fox squirrels than if they are the red or flying variety. The former will likely face you down as soon as you invade their territory, while the latter will discreetly take advantage of any available hiding place rather than challenge so large an intruder.

### From Bad to Worse

Squirrels are loaded with parasites, inside and out. Infestations of fleas, which happily make the transition from squirrel to human hosts, are common, as are plagues of lice and mites. Fox squirrels in some areas are underweight and have missing patches of hair due to mites that cause a mangelike case of skin infection. Squirrels also carry transmittable diseases, including encephalitis and typhus. In some cases, the fleas that are along for the ride have been found to carry sylvatic plague. Finally, when resident squirrels expire, their remains will rot and emit a disgusting odor for several weeks. This in turn attracts other rodents, flies, and creepy things to your attic. On a lone positive note, although squirrels have been known to carry rabies, they seldom escape the attack of any rabid beast long enough to pass it along.

## *High-wire Acts*

One of the purest of unexpected joys to any city dweller is the sudden burst of glee felt upon spying an urban squirrel confidently crossing the street. As surely as if it had its own crossing guard, the squirrel scampers safely across, paying the traffic no heed, for it has its own private crosswalk high above the rabble: Utility wires make the perfect squirrel highway. Unless, of course, the squirrel comes into contact with a transformer. Electricity and squirrels don't mix.

Wire-walking squirrels are typically not suicidal. Some are just intent on disrupting our communication systems or, at the very least, messing with our cable TV. Squirrels gnawing through the exterior protective sheath of aerial cables jeopardize the integrity of the system by allowing water and sunlight to enter and damage the interior wires.

## *Ground Zero*

Digging and chewing at ground level can have serious consequences. From chipmunks to woodchucks, almost any member of the squirrel family will find some way to get into trouble without ever setting paw in your house or garden, or upon your overhead wires. Their prime directives to dig, chew, and swallow won't allow them to leave any stone unturned. Or maybe they're just looking for some good dirt on us.

Anything a squirrel can get into is fair game as far as the animal is concerned. Regardless of whether they're interested in the contents, squirrels seem to delight in digging out whatever is inside a flowerpot. Dog or cat food left outside is also an open invitation. Most of these pet foods are grain based and high in protein. The higher-priced varieties throw in a good amount of quality fat to make them more palatable. What more could a squirrel ask for? Hmm . . . perhaps a little variety, such as the contents of a garbage can. No telling what wonders a persistent squirrel might uncover if it just keeps its nose busy and lets its digging claws follow. Just as in attics and on utility poles, low-level wiring is always a target. Next time the TV goes out, check the wire entry hole to the premises before you automatically make that call to the cable guy. Likewise, plastic or rubber hoses seem to be favored as squirrel floss. Tubing on gas grills, perhaps because of the lingering summer scents, is routinely gnawed through. Anyone who has collected maple sap knows that the lines are especially vulnerable to squirrels, sweet on the flavor of fresh syrup.

But chewing is the least-damaging hobby of a determined squirrel. Chipmunks, ground squirrels, and sometimes woodchucks can severely undermine the footings or foundations of outbuildings. They have also been known to tumble stacks of firewood and, more ominously, damage the basement or foundation of your home. Diligent digging near structures always spells serious trouble!

## *When Squirrels Attack*

In the minds of some people, cute little squirrels could never possibly be considered a pest or nuisance. Don't be fooled. Even college kids have learned firsthand that squirrels are ornery.

Campuses across the United States and Canada afford some of the finest living arrangements to be found in squirreldom. Stately, mature shade trees, lush, well-kept grounds, and people who are young and innocent enough to still be optimistic about the world of squirrels. People, in other words, who are willing to share their French fries. On many campuses, squirrels have earned quite a reputation. The squirrelly phenomenon even has its own Web site (www.scarysquirrel.org).

Squirrels delight in digging up potted plants, sometimes sowing their own seeds.

Not long ago, a column in South Florida University's newspaper, *The Oracle,* pleaded with fellow students and administrators to stop the madness. At issue was the fact that, having come to expect handouts, the squirrels were bypassing the niceties of acting cute and waiting for a treat in favor of a more direct approach, blatantly mugging passersby and demanding a handout. Elsewhere, a couple of Harvard students perfected a way of offering squirrels a snack without risking body parts. They swing a treat at the end of a string — squirrel fishing.

If young adults are at risk, so too are children, who may reach out on a whim to "pet the nice squirrel." The elderly, who quietly enjoy the antics of squirrels at the park or in their backyards, should also be careful. Everyone should be warned never to attempt to touch a squirrel. Squirrels are wild animals, and they bite.

Once squirrels, especially the gray and fox species, lose their fear of humans and discover that we are essentially just big, mobile food dispensers, they can become quite aggressive. And they will bite the hand that feeds them without a second thought. Those razor-sharp teeth, honed to a fine cutting edge on tough nuts and miles of television cable, can cause serious damage.

Harvard students offer treats to aggressive squirrels from a safe distance.

## *Environmental Impact*

In some areas squirrels have had a significant impact on the environment around them — and not just by planting trees. Trees, in fact, also *suffer* at the paws and teeth of squirrels. Red and fox squirrels have a serious impact on timberlands from the California coast to the southeastern corner of the United States. They are very fond of the bark of the dawn redwood tree, which they strip and use to line their nests. Any tree is at risk, but younger trees that have thin bark are most susceptible. Trees that are stripped by squirrels grow more slowly, and those on which the bark has been stripped clear around the trunk, or *girdled,* die.

Pine trees frequently suffer indirect damage from squirrel activity. The overzealous varmints rip the green cones from the stem, leaving open wounds that are susceptible to insect damage and disease. The resulting dead branch tips, or *flagging,* occurs most often in years when the squirrels have little to eat but pinecones.

Squirrels also have an effect on other creatures. Franklin's Ground Squirrels, for example, are known to destroy the nests of pheasants and ducks, killing and devouring the occupants in the process. One study that examined squirrel activity along the southern edge of Lake Manitoba concluded that 19 percent of the duck nests in the area had been destroyed by squirrels. The California Ground Squirrel's taste for quail eggs has long been suspected of limiting quail populations in some parts of the West.

A less gruesome aspect of the impact squirrels can have on their environment is illustrated by the contribution that ground squirrels and prairie dogs make to the evolution of grasses in their range. Continual cropping of longer-stemmed varieties of grasses opens the area up to reseeding by different varieties. Over time, entire stretches of open grassland are replanted. Such was the case on the American prairies during the heyday of the prairie dog towns.

CHAPTER 5

# Defending Your Bird Feeder

**For some readers,** this chapter gets down to the main reason you picked up this book in the first place. Or at least what you originally *thought* was your reason. Having since learned what wonders and threats squirrels really are, perhaps you've decided that there is more there to contend with than just a little wasted birdseed. Or not. At any rate, here are some proven methods for keeping squirrels out of your bird feeder.

## *Some Commonsense Solutions*

When it comes to guarding the bird feeder, there are only so many options. Place it out of squirrels' range, make it tricky or impossible to get to, or make the contents not worth the effort.

## Avoid Launching Sites

Birds can fly; squirrels can't. This ought to give our feathered friends an advantage in reaching those seed-bearing outposts we optimistically call *bird* feeders. Of course, squirrels are smart and tough, with a tremendous capacity to learn from both their mistakes and their successes. But since every species of squirrel (with the notable exception of the flying varmints) must leap, climb, or descend to the bird feeder, there is a physical limit as to how far they can go to reach it.

In order to give yourself a fair shot at winning this battle, think location, location, location. Find a spot for your bird feeder that has good viewing potential, well away (at least 20 feet [6.1 m]) from your house. No sense inviting the bushy-tailed home wreckers right up to your doorstep. Try to find a spot for the feeder that is readily visible from at least one window, preferably one near which you normally spend a fair amount of time. This will add to your enjoyment of watching the birds, as well as simplify your surveillance of squirrels.

Knowing your squirrels, their methods, and their routes to your bird feeder makes a world of difference in figuring out the best ways to stop them, or at least slow them down. Knowing what you're up against is crucial to success. It may be that you don't know the exact capabilities of your own squirrels. (Yet. You will. It's only a matter of time.) Don't despair. For the most part, there are limits to what they can do.

> "Don't underestimate the power of [squirrel] trial and error. Squirrels are going to spend more time trying to get to your birdseed than you want to spend trying to keep them out."
>
> – Cornell Ornithologist Todd Culver, *Cornell Chronicle*, 1992

The average squirrel, if indeed such a critter exists, is limited in its bird-feeder exploits by how far its little leg muscles can propel it. Most squirrels can spring a good

4 feet (1.2 m) straight up from the ground, which jeopardizes most pole-mounted feeders that are placed within handy reach of the designated feeder filler. Therefore, pole-mounted feeders need be more than 4 feet (1.2 m) off the ground. Bear in mind that some squirrels can jump higher than this, so consider 4 feet (1.2 m) the minimum. (And check out Protecting Post-mounted Feeders on page 103 for more suggestions.)

This same average squirrel can launch itself from 5 to 10 feet (1.5–3.1 m) horizontally, which means that if it has any kind of a perch within range of the bird feeder, it'll be dining on birdseed before you can blink. The standard rule is to be sure the feeder is placed no closer than 10 feet (3.1 m) to any potential horizontal perch. Unfortunately, this solution yields a further problem: Many birds are a tad shy and are more apt to visit your feeder if there is at least a little cover nearby. And low-lying cover is a definite no-no, because it affords protection for domestic cats and other predators.

One suggestion is to provide a thorny, old-fashioned rosebush near the feeder as a quick escape for nervous birds.

Squirrels can leap onto pole-mounted feeders from up to 10 feet (3.1 m) away.

## Feeder Placement Is Critical

Be sure to place your bird feeder:

- Where you can view it from indoors
- Away from the house
- More than 4 feet (1.2 m) high
- At least 8 to 10 feet (2.4–3.1 m) from any lateral jumping-off point
- At least 12 feet (3.7 m) from overhead jumping-off point
- With cover within 12 to 20 feet (3.7–6.1 m) for nervous birds
- With *no* nearby ground cover to hide would-be predators

(The thorns will dissuade predators from using the shrubs as camouflage.) Old-fashioned roses are the best choice, because they are bushier and require less maintenance than do the stingy stems of most modern hybrid teas. Ask your local nursery which thorny varieties do best in your area.

The final consideration in placing a bird feeder out of reach of that average squirrel is vulnerability from above. Watch out for utility lines, overhangs, and eaves. If there is a ledge, edge, limb, or clothesline within 12 feet (3.7 m) above the bird feeder, consider the feeder accessible. Forget any quaint ideas about hanging a bird feeder from a tree limb. Anything hanging from a tree is automatically considered squirrel property and will be dealt with accordingly. Trim back overhanging tree limbs when stationing a bird feeder nearby, or better yet, move the feeder to a more open area. In short, when placing a bird feeder, look up, look down, look all around for potential sites from which a hurtling squirrel might alight upon it.

If all else fails and you can't eliminate all potential launch sites, consider sabotaging them. Cover them with Nixalite, slather them with repellents, place obstructions in the way, coat them with unappealing substances, or tie the dog to them. (See chapter 6 for a detailed discussion of squirrel repellants.)

## Change the Menu

Since you've already had to admit that you're dealing with an intelligent, persistent, brazen foe that really doesn't have much else to do, perhaps it's best to consider the situation from the squirrel's perspective: It is making a choice to raid your bird feeder, for you made it an offer it couldn't refuse. So change the offer.

Squirrels learn through trial, error, and *reward* that all the trouble they go through to get to that well-stocked bird feeder is worthwhile. As wild creatures, their instincts for self-preservation tell them to take the easiest route to food, expend the least amount of energy, and avoid peril. And to the squirrel, bird feeders are good pickin's compared to what nature has to offer.

Think, for a moment, like a squirrel. "*Mmmm* sunflower seeds. . . . *Mmmm* peanut hearts. . . . *Mmmm* pecans." You make a mad dash through the open yard, past the many bare spots in the lawn from which you have already excavated seeds and nuts from previous caches. You dodge the dog, dart up a nearby tree, and sprint out onto a limb that wobbles beneath your weight. There, you cling tenaciously to the smallest end of the branch as it bounces up and down, carefully calculate your trajectory, and, with every ounce of strength your hind legs can muster, launch yourself from the branch just as it hits the apex of an upswing. You are airborne! With only a split second to adjust for a perfect four-point landing, you twist in midair and . . . plunk! You've made it. You're on the feeder. Oh, happy day! Reach in there, big fella, and scoop up a handful of your hard-earned reward.

### Playful at Heart

Squirrels seem to enjoy the challenge of getting to the food reward on your feeder. Once they've mastered something, such as hanging by one foot from an overhead baffle to reach for sunflower seeds, they appear to relish doing it again to prove it wasn't a fluke. Bird feeders offer food and fun.

"P-tooey! What's this? This stuff is awful. Where's the sunflower seeds?" Dig some more; cast the bitter-tasting safflower seeds aside. "Where's the peanut hearts?" Desperately fling the thistle seeds to the ground. "Nooooooo!"

Although it may take the varmint a few visits to realize what's in store, no squirrel in its right mind will work for a reward it doesn't like. Consider changing the menu to those (few) items that squirrels just don't think are worth the effort.

The downside to limiting the offerings at the bird feeder is that it also limits the potential avian visitors. To effectively discourage squirrels, the feeders must be stocked solely with things they find unappetizing, attracting fewer varieties of birds in the process. Still, cardinals, doves, chickadees, titmice, evening grosbeaks, nuthatches, and house finches will come for safflower seeds and, as a bonus to some bird enthusiasts, most blackbirds won't (red-winged blackbirds being the exception). Goldfinches actually *prefer* thistle seeds, and house finches, pine siskins, mourning doves, and juncos will politely accept this offering as well.

While both squirrels and birds go for suet blocks loaded with bits of grain or seeds, it's the garnishes the squirrels are after. Few squirrels are desperate enough to chow down on pure, raw beef fat. Opt for plain suet blocks to discourage them.

An alternative to changing the menu is to spice it up. Squirrel Away, a taste repellent, is derived from capsaicin, which is the chemical that makes hot peppers hot. This is the same chemical that is used to ward off grizzly bears. Spritzed onto bird food, capsaicin burns the squirrel's sensitive mucous membranes and teaches it that the food at this joint comes with a dose of pain. Better yet, capsaicin

### Stuff Squirrels Don't Like

- Pure raw or commercial beef suet
- Safflower seed
- Thistle seeds or niger*

* Not related to common garden thistles, this tropical seed is heat-sterilized and will not sprout in the yard.

poses no threat to birds, as their beaks are not affected by the stuff. A poll taken by *Wild Bird* magazine reported Squirrel Away to be effective about 50 percent of the time. Those readers who responded to the poll offered mixed opinions. Some reported great success, while others found that the squirrels seemed to just shrug off the discomfort.

## Git Along, Little Critters

Extreme changes in environment will convince squirrels to relocate. Clear-cut logging is one example; paving over acres of land to create parking lots is another. Many years ago, Native Americans made quick meals of prairie dogs and other ground squirrels by flooding them out of their burrows. But some slightly less drastic options are also effective and may be more appealing to the homeowner. Keeping trees trimmed and thinned is a good start. Don't provide squirrels with food, water, or shelter. This means feeding your own critters indoors and forgoing the bird feeder, at least for a while. Don't leave anything that may be considered squirrel food out where they can go for it. Make sure garbage cans are not accessible to the little thieves; check that lids fit securely. If you have a garden pond, cover it at night. Remove any debris piles, and enclose your compost heap.

> **Squirrelly Facts**
>
> Herr Kemel, a German inventor, used squirrel fur to make a fine brush for detailed artwork. The brush is commonly called the "camel's-hair brush."

The truth is that on the small scale most homeowners and bird enthusiasts have to work with, there is little you can do by way of habitat modification to steer squirrels clear of your bird feeders. However, there is a *lot* you can do to avoid asking for trouble. The key rule is not to invite them in the first place.

## Make a Better Offer

Most veteran bird feeders will tell you that the best way to keep squirrels out of the birdseed is to distract them with easier pickin's. This may sound like a defeat, but sometimes you have to concede the battle to win the war.

One longtime bird-feeding enthusiast suggested simply putting up extra bird feeders and keeping them all well stocked. "As long as there is enough for everyone, they will all share," she assured me. I'm not sure I'm buying it, and I'm definitely not convinced I want to go to that much extra trouble and expense. However, in cases of mild squirrel pillage, it might be the easiest and least stressful solution.

Squirrels are opportunists, which is one key to their ability to survive in the wild. If you offer a squirrel an alternate food source that it can reach with minimal effort, perhaps, just perhaps, it will leave your bird food for the birds. Consider erecting squirrel feeders as a means of keeping the hairy little thieves preoccupied and satisfied. If this sounds like bribery, that's because it is. Just remember to locate the squirrel feeders far away from your bird feeders.

There are several types of commercial squirrel feeders available. Since this is basically a peace offering, stick with the principle that simpler is better. Unless you are prepared for revenge in the form of chewing, dislodging, or outright destruction, refrain from trying to get squirrels to perform for their food. Feeders that require squirrels to open a lid,

### Stuff Squirrels Really Like

When you resort to catering to squirrels to keep them from your bird feeders, remember, this is bribery, so don't go overboard on expensive treats like pecans when the cheaper versions, such as peanuts, will do. The following are some popular, inexpensive squirrel treats:

- Corn
- Fresh or dried fruits
- Peanut butter
- Peanuts, roasted and unsalted
- Sunflower seeds

move a lever, or actively work for a reward can backfire. Frustrated squirrels resort to either returning to the bird feeders (which they have already figured out how to beat) or engaging in the activity they know best — chewing. Squirrel feeders with clear plastic panels that show the food, yet don't easily dispense, appear to be the most frustrating. On the other hand, feeders such as the rotating corncob wheel seem to engage the squirrels as much with their entertainment value as their treats.

A simple shelf, or a platform feeder with a roof to protect the varmints from the elements, are the paws-down favorites among squirrels. Some commercial shelf-type feeders come with a spike for jamming dried corncobs onto them. Affix the feeders in a tree at a height that allows the squirrels to feel safe, typically from 8 to 20 feet (2.4–6.1 m) up, and they will munch unmolested. Who knows, you might even enjoy watching them. Be sure to mount squirrel feeders a respectable distance from your house, lest you inadvertently offer the critters lodging as well as meals. Consider 20 to 30 feet (6.1–9.2 m) a minimum safe distance.

### *Squirrelly Recipes*

Those kind folks who go out of their way to feed squirrels have come up with two ready-mix formulas that are not only tasty but nutritious as well.

Whether you agree that intentionally feeding squirrels is a good idea depends on the circumstances and your personal perspective. Supplementing the food of any wild animal alters its life substantially. Well-fed squirrels are more likely to bear more young, resulting in more squirrels to feed. Lots of squirrels need lots of nesting space, possibly resulting in them seeking refuge in places you'd rather they didn't, such as your attic. Of course, you can always resort to installing squirrel nestboxes to accommodate the extra squirrels you help create. Sound like an unending cycle?

## Squirrel Trail Mix

1 part commercial birdseed mix (also attracts birds to clean up the mess the squirrels will inevitably leave)
2 parts black sunflower seeds
1 part cracked corn
1 part whole peanuts, in the shell
1 part dry dog food

Mix and serve.

## Squirrel Dough

1 part chunky peanut butter
1 part dried fruit
1 part shortening
4 parts coarse cornmeal

Mix the ingredients together to form a dough. Slice the dough into chunks and slather it onto pinecones or directly onto tree bark. Keep unused portion refrigerated.

On the other hand, if you are already sacrificing bird food to squirrels, you are in effect already supplementing their wild diet, albeit unintentionally. It doesn't matter to the squirrels: Either way, they eat.

## *Protecting Post-mounted Feeders*

So your yard isn't big enough to hold a spot far enough in the open that squirrels will leave your bird feeder alone? Here are some tips from those who have gone to battle and won for protecting post-mounted bird feeders.

### Greased Poles

Many bird-feeding veterans suggest greasing the support pole that holds up the feeder, thus preventing squirrels from climbing up for a meal. Although this strategy does work (for a while, anyway), there are some immediate drawbacks.

Axle grease, the compound most often recommended for slicking up bird-feeder poles, is probably the least environmentally friendly. For starters, it results in dirty, greasy squirrels, and who knows what they'll jump on next. Second, it can be harmful to the varmints as they try to clean the gunk off their fur. Ingested axle grease is not healthy. C'mon, we want to discourage them, not poison them. Third, it gets tacky in cold weather — the very time when the birds are most likely to be depending on your feeder and, coincidentally, when the squirrels are most likely to have little else to entertain them. It's no fun, and somewhat humiliating, to watch a squirrel trot up a sticky pole, laughing at your feeble efforts.

Alternate greasy substances, from petroleum jelly to shortening to (yuck) animal fat, have all been tried. The results, unfortunately, are roughly the same — failure. Furthermore, some greasy products, such as bacon grease, may

### Metal Mouth

The magic word with squirrels is *metal*. It's hard to grip, and they can't chew through it. Whenever possible, use metal mounting posts or brackets and store any birdseed kept outdoors in metal canisters or garbage cans. Don't worry about the metal hurting birds during periods of cold weather. They don't have sweat glands in their feet, so they won't stick to metal perches, and their tongues have a covering that protects them as well.

seem like a good idea at the time, but they wind up attracting every dog, cat, and rat in the neighborhood. More aromatic greases, such as petroleum jelly mixed with menthol oil (or smelly, medicated oil-based products like Bengay or Vicks Vapo-Rub), are more effective, especially at first. Not only do they offend the squirrel's sense of smell, but they also mask the aroma of food that is emanating from the feeder.

## Polybutenes

These are sticky substances that squirrels find repugnant to one degree or another. A neat and tidy tip is to wrap the post with duct tape or paper tape prior to coating it with this gunky stuff. Polybutenes are available in a range of forms, from slippery liquids to extremely viscous, tacky pastes. They can be dabbed on with a rag or spread on with a wooden spatula or trowel.

## 4-The-Birds

This is a sticky substance that irritates the skin. Pasted along the length of the post, it quickly discourages squirrels from coming into contact with it. It weathers well and can last up to a year. Available in liquid spray form and clear gel, it's easy to apply and made-to-order for discouraging "roosting," sitting, climbing, or any other squirrelly means of trespassing.

## Teflon Coating

Spray-on Teflon coating is the high-tech version of a greased metal pole. The coating is slippery, won't come off on the squirrel, and lasts for several rains. A huge disadvantage, however, is that most teflon applications must be "cured" at 700°F (371°C) or higher.

## A Spiny Defense

At lease one intrepid squirrel antagonist I know has had success covering the feeder and wrapping parts of the support pole with Nixalite strips. These are flexible, thin metal strips covered with needle-sharp points, designed to dissuade critters from alighting or gaining a foothold. The strips will bend to fit on or around virtually any feeder or pole. If squirrels are bold enough to test the metal, they will quickly get the point. The downside is that this product makes your bird feeder look like a war zone. It can also pose a hazard to pets, children, and sometimes even birds.

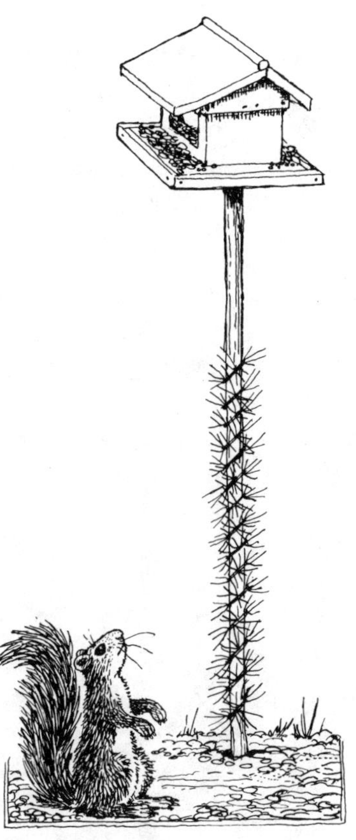

The needlelike spines of Nixalite stripping send a "paws off" message to snooping squirrels.

## Paint It Red

At least one study suggests that squirrels don't like the color red. Being natural enemies of certain types of red-headed woodpeckers, whose niche in the wild they energetically compete for, perhaps this is an adaptation to keep them out of trouble. Companies that manufacture equipment to keep squirrels away from transformers and other high-voltage electrical components paint their products red in an effort to exploit every possible squirrel-deterring factor. Conceding that every little bit helps, consider adding red into *your* backyard color scheme.

## Baffle 'em

The good news is that properly installed squirrel baffles, whether store bought or homemade, do deter squirrels. (There are several marketed for use on metal pole supports.) The bad news is that a baffle installed to protect from only one direction, such as from below, does not eliminate the need to be aware of squirrels coming in from different angles. One such baffle that has received good customer ratings is the ERVA Squirrel Baffle. This product is 6 inches (15.2 cm) in diameter, 14½ inches (36.8 cm) long, and built to thread over a metal pipe. A do-it-yourself version might be made from a cylinder of sheet metal that has been fitted with ends and threaded over the pipe.

A saucerlike baffle tips when squirrels try to use it as a stepping-stone to your feeder.

Some successful squirrel baffles look like an upside-down saucer. A tip from someone who has consistently had good luck with hers is to mount the baffle about halfway up an 8-foot-high (2.4 m) post by resting it on a stop, which keeps the baffle from sliding to the base of the pole. Secure the stop, but not the baffle. This way, every time a squirrel makes a grab for the baffle, the device tips and promptly deposits the intruder back on the ground.

## Spinning PVC

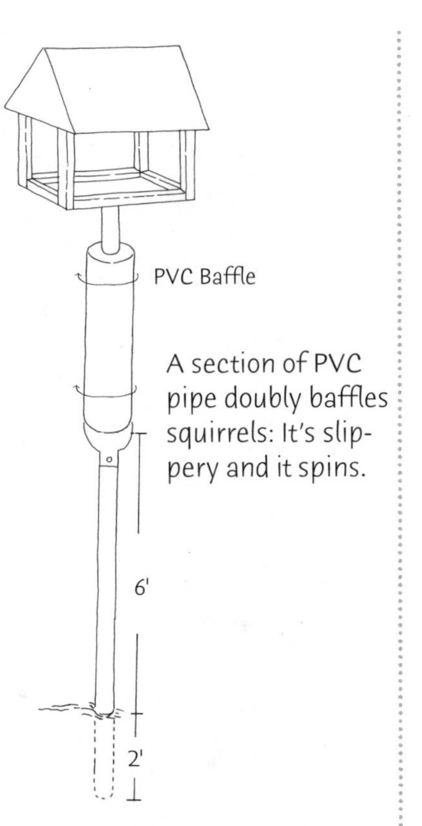

A section of PVC pipe doubly baffles squirrels: It's slippery and it spins.

Another option is to fit one or more 18-inch (45.7 cm) sections of plastic or PVC pipe over a metal post. The spinning pipe(s) make it difficult for even the most talented squirrels to get a grip. Alternately, try using the plastic pipe for the bird-feeder post itself. Equipped with a flange at the top just beneath the feeder, the combination of a slippery post and a barricaded top will keep out most squirrels. The trick is to use smooth, slick pipe. Allowing the pipe to become scratched or roughed up will make it much more climbable and defeat your purpose.

## Wooden Posts

Wooden posts make attractive and sturdy bird-feeder supports. They also make for exceedingly easy climbing by a hungry squirrel. If you are crafty, you can fabricate a sheet-metal or acrylic barrier as explained above. Again, the trick is to place the post and feeder so that there is no other way for the squirrel to access the feeder, and to place the baffle at least 4 feet (1.2 m) from the ground. The general rule of thumb seems to be that the more unsightly the baffle, the more effective it is.

## Electroshock Therapy

Sometimes you have to appreciate those few truly demented souls who seek to foil squirrels at their own level. One such individual has found a way to deter squirrels from even attempting to accost his feeder. He electrified the pole. Shocking!

A safe way to charge a bird-feeder pole is to attach it to an electric fence charger, which is specifically made for charging livestock fences. You can either electrify an entire metal post (insulated from the ground), or electrify a long strand of electric fence wire (be sure to wrap the wire around a plastic post for insulation). Birds lighting on the feeder are not grounded and eat undisturbed, but let a squirrel (or cat, or other unauthorized critter) touch the pole while in contact with the ground and, well, it'll get a charge out of it. Just be sure to disconnect the charger whenever you refill the feeder, or you will, too.

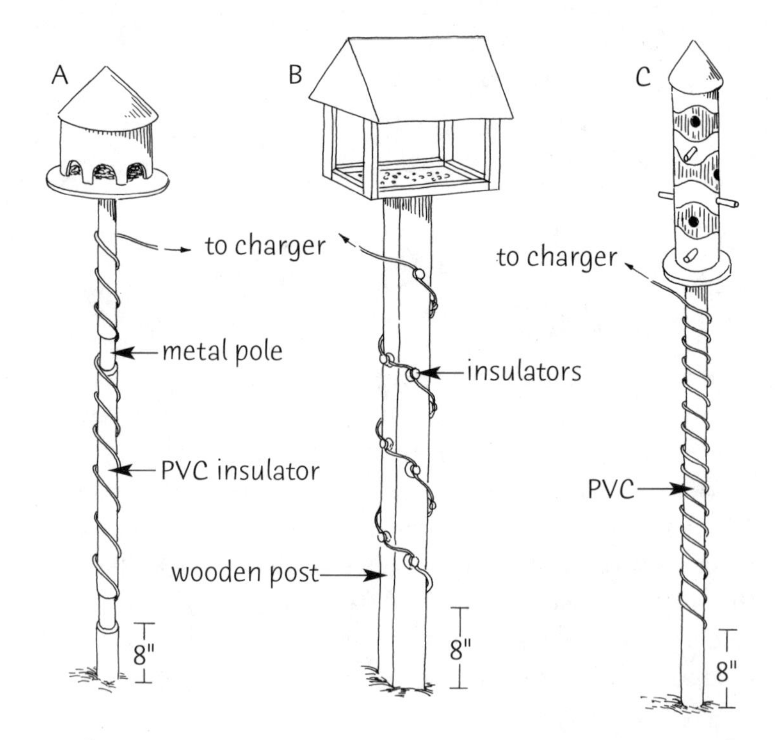


(A) Fit a metal pole with sections of PVC pipe to act as insulation between the electrified wire and the conductive metal.
(B) For a wooden post, attach plastic insulators along the length of the wire at intervals sufficient to keep the wire from touching the post.
(C) For a PVC post, simply wrap electric fence wire around the post.

After the squirrels have been trained to avoid the bird feeder, you may want to disconnect the charger. Bear in mind, however, that livestock always seem to know when the fence is on or off, sticking their head through or walking over the fence when it is not charged. Expect no less from a squirrel.

## *Protecting Hanging Bird Feeders*

The most common alternative to mounting your bird feeders on a post is to hang them from a wire. For many people, this is the most convenient solution. It also opens up some creative ways to deal with persistent squirrels. The most important general rules to remember involve proper spacing and not using a chewable substance to hang the feeders. These include cotton, jute or nylon, twine, rope, and plastic. Stick with wire or metal chain.

### Hang 'em High and Wide

Squirrels are phenomenal tightrope walkers. It takes more than a cable, wire, or rope strung across an expanse to stop them. A long, thin wire that crosses the yard 20 to 30 feet (6.1–9.2 m) from the ground, however, makes a great squirrel stopper. Hang feeders from the wire at intervals, making sure that the secondary wires from which the feeders actually hang are long enough to allow you to reach the bird feeders for refilling (see below). Depending on the length of the wire and the size of your yard, several feeders can be strung along such a line to cater to different species of birds.

### Second-storey Ingenuity

For folks with a two-storey house, or who live in a second-storey or higher apartment, the way to hang bird feeders out of the reach of squirrels is just a pulley away. Set up a pulley wire from a point opposite a favorite window, and provide the birds with a high-wire act they can't resist. Simply pull in the feeder for refilling. At such a height and with no point of access, disappointed squirrels will be forced to admit defeat.

A second-storey pulley system makes filling feeders easy while denying access to squirrels.

## Hang 'em Low and Close

For those who would rather not try the high-wire act as mentioned above, there is a simple, down-to-earth (well, nearly) alternative. Locate two points in your yard from which you can string a thin stretch of wire. These could be trees, the side of a building, or even the edge of the clothesline. Make sure they are at least 16 feet (4.9 m) apart and high enough that squirrels can't jump up to them (and that you won't accidently walk into the line once it's up).

Once you have selected an appropriate spot, measure the distance between the two points and cut a length of wire to match, leaving enough extra to tie off at the ends. Before you hang the wire, pierce a hole through the bottom center of 8 to 10 sodapop bottles and through the center of each bottle cap. Push the strand of wire through the bottles until you have half the bottles strung, then affix a loop or ring from which to hold the bird feeder. Finish by stringing the remaining bottles. (For variations on the sodapop bottle setup, use record albums, sections of PVC pipe, coffee cans, or old plastic containers such as margarine tubs.)

If you think this contraption looks weird, that's nothing compared to how cumbersome it is to actually hang the bottle-bedecked wire. Tie one end off at your first predetermined point and drag the line, bottles attached, to the other point. Tie it in place there. (Ignore the squirrels in the tree that are sitting there laughing at you. They don't know what they're up against.) You may want to wire in a turnbuckle at one end to keep the line taut, because the weight may cause it to sag over time. Use a stepladder, if necessary, to attach the bird feeder in the center, and stand back. As soon as a squirrel sets foot on one of the bottles, it'll get spun right back to earth!

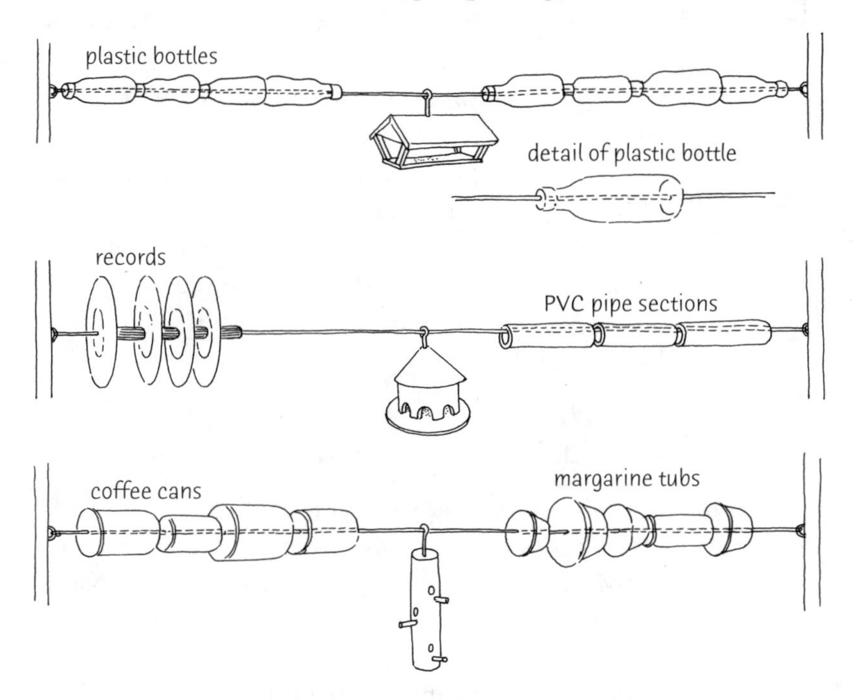


Build a daunting obstacle course around your hanging feeder with plastic bottles, records, margarine tubs, and other household items.

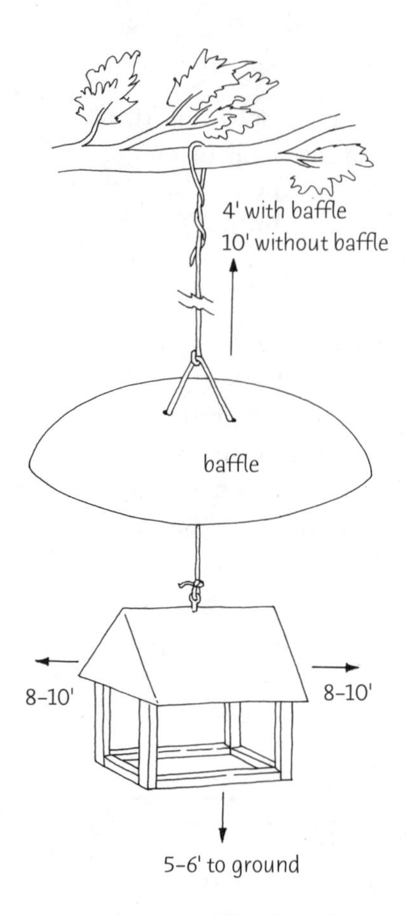


An overhead baffle placed at least 4 feet (1.2 m) down a hanger halts skulking squirrels and tips them to the ground.

## Baffle 'em from Above

Umbrella-like overhead baffles are quite effective at keeping squirrels from bird feeders that are hung on wires. Just be sure the baffle is big and not held too firmly in place. A baffle that is too small will actually allow the squirrels to reach around or, in an apparent attempt to show off how limber its toes are, to hang by its back feet and reach into the feeder with its front feet. By attaching the baffle on a stop so that it tips back and forth, you make the obstruction a precarious perch as well.

Also, be careful to position the baffle at least 4 feet (1.2 m) down a hanging wire or cable: Squirrels find ways around those that are positioned too close to a jumping-off point. As for positioning in general, be certain there is no accessible area from which a squirrel can jump to the feeder, either from the ground or from the sides.

### Not So Smart: Hanging Feeders from Tree Limbs

I admit it. That's where mine's hanging. And in general it's a terrible idea. First of all, trees are squirrel country, no way around it. Anything hung from a limb is their property, plain and simple. Second, I managed to hang mine from a pine tree, which has lots of limbs from which squirrels can dangle. A good rule of thumb is to find a deciduous tree from which to suspend the feeder. It will allow much clearer viewing of the bird feeder in winter when the limbs are bare, while affording the squirrels less access.

If you're hanging the feeders from metal wire, always use a tree guard, or thread the wire through a section of rubber garden hose to protect the tree bark. As with wire-hung feeders that use baffles, hang the feeder at least 4 feet (1.2 m) beneath the branch and at least 4 feet (1.2 m) from the ground. If you can find, and reach, a limb 20 feet (6.1 m) above the ground, attach a wire about 12 feet (3.7 m) long from which to suspend the feeder. This way, the feeder should hang below the squirrels' jump-down-to range and above their jump-up-to range.

## Other Squirrel-proofing Measures

Not satisfied with baffles, huh? Think you need to rid the area of squirrels to keep ahead of their ravenous raids? Here are more solutions.

### Dogged Deterrents

I love dogs. Mine is a household that will never be without them, usually *big* dogs. Any dog you have to bend

### Baffling Thoughts

Lots of well-made commercial squirrel baffles are on the market, and just about any garden supply store, specialty bird-feeding shop, or bird catalog will carry an assortment of baffles of varying designs and materials. But if you're a handy person or just like to do things the hard way, you can make your own. Here are some suggestions, from the sensible to the far out to the just plain ugly:

Overhead baffles:

- Cut to measure, heat, and fashion Plexiglas.
- Cut to measure, bend, and fabricate with sheet metal and screws.
- Use an old metal garbage can lid.

Climbing baffles:

- Cut sections of black plastic pipe.
- Use an old section of stovepipe, with ends covered.
- Thread two or three 5-pound (2.3 kg) coffee cans over the mounting pole by drilling a hole in the base of each and cutting a hole in the lid. Position with the lids together.

over to pet can't really be much of a dog, can it, now? Big dogs scare away the Avon lady, trick-or-treaters, various prize patrols, and, on a good day, squirrels.

## Ultrasonic Noise

Wouldn't it be lovely if you could just flip a switch and the squirrels would instantly leave your bird feeder alone? That's the theory behind ultrasonic noise machines.

First, a little history on these gadgets. Ultrasonic noise devices were originally intended for use in warehouses, where the obstructions for the sound waves were few, and the rodents were many. Speakers from the units emitted a stream of noise that was too high in frequency for people to hear, but well within the range of sounds that small animals can detect. In fact, it was believed that the range of the speakers was roughly equivalent to the range of high-pitched distress calls for the very rodents the contraptions were intended for. What was silent to humans was a scream to pests.

And they worked. They worked so well, in fact, that it wasn't long before someone thought of marketing them for home use. The problem was, the industrial-strength models were large and expensive. So changes were made to scale them down for home and garden use, and that's when the problems began. Many of the smaller models released for home use just weren't up to snuff.

Another concern often expressed regarding the use of ultrasonic noisemakers is that the rodents will eventually get used to the sound. This is possible, but if you have one of these machines and the squir-

### Hummingbird Feeders

Squirrels will go to some incredible lengths to get a satisfying slurp of the sweet nectar you put out for hummingbirds. The same squirrel-proof strategies that are recommended for seed feeders will prevent the eager varmints from sucking the nectar feeder dry.

rels seem to ignore it, more than likely the problem is that the critters aren't hearing the sound. Let me explain. Sound waves travel in a straight line; they don't bend, go over, under, or through. If they hit an obstruction, they stop. Thus, if the machine is placed in an area where the sound waves are blocked, the critters won't get an earful. The user, of course, thinks, "Shucks, the darned thing doesn't work." There may be some credence, however, to the idea that the varmints get used to the high-pitched screams. Most of us get used to our kids.

**Nutty Facts**

When squirrels fall from great heights, they use their tails as parachutes to slow their descent.

For use against squirrels at the bird feeder, an outdoor ultrasonic device may be worth the experiment. Look for a model that has metal, not paper, speakers to provide optimal sound quality. Be sure you order one that is large enough for the area you need to cover. Manufacturers state the area coverage with each model. Also, try to get one with a money-back guarantee in case the experiment fails. Then plug it in, aim it at the bird feeder, and see what happens.

## Repellents

There is a plethora of products that claim to repel squirrels. Known as "area repellents" because the smell is supposed to keep unwanted critters out of a treated area, most are not effective against squirrels bent on gorging at your bird feeder.

One way to use a scent repellent to keep squirrels from pestering the feeder is to slather it on the pole or on baffles on the side that the critters most often frequent. One repellent that is reported to work is commercially marketed fox urine, which is supposed to scare the squirrels into believing a predator is lurking nearby waiting to devour them. Used kitty litter scattered near the base of the post is a do-it-yourself version of the same theory that offers mixed results.

Other products act on similar principles. Rid-A-Critter is an area repellent consisting of naphthalene, the same material that puts the stink in mothballs. Squirrels don't like the aroma, so sprinkling it around the base of the feeder may keep them at bay. However, if the reward in the feeder is great enough, the squirrels will eventually get used to the smell and ignore it, especially outdoors, where the vapors are not nearly as concentrated as in an indoor setting. Blood meal is somewhat effective in deterring determined squirrels. Working some into the soil at the base of the pole may get them to retreat. If not, it will at the very least improve your soil. These scents won't work forever, especially not in rainy weather. They may, however, buy you some time as you search for other cures.

### Biting the Hand That Feeds Them

If squirrels have developed a taste for the feeders themselves, as is often the case with cedar-sided feeders, consider coating the outsides of them with a contact repellent such as Ropel. It tastes horrible and at the very least will discourage the scoundrels from gnawing at the wood.

## *Broadening Your Scope*

In the war against squirrels, sometimes it helps to take a step back and look at the big picture. If squirrels are raiding your feeder, it makes sense to expand your squirrel-proofing efforts and take steps to keep the varmints from your entire yard. If they aren't in your yard, they won't raid your feeder. Chapter 6 explores the various lines of defense you can establish to create a squirrel-free zone in your yard and in your home.

CHAPTER 6

# Protecting Your Garden and House

**As we have seen,** squirrels have covered just about every angle in their multifaceted plan of attack against humans. They have come at us from the air, the land, and underground. Stories abound of squirrels swimming rivers and lakes in their unending mission to conquer the world, or at least our yards and gardens. But we should not admit defeat. Believe it or not, there are steps you can take to keep even the most persistent squirrel out of your yard, garden, and house. This chapter provides plans for counterattack. Fence in the garden and call out the dogs; repel with smell and cage 'em if you can.

# *A Squirrel-proof Garden*

Next to raiding bird feeders, one of the most common complaints about squirrels concerns their penchant for harvesting crops they didn't plant. In other words, squirrels of all varieties have a knack for using *your* garden to satisfy *their* hunger.

Since squirrel damage in the garden stems either from digging, eating, or both, there are two simple ways to try to approach the problem: Stop them from digging, or stop them from eating.

## Creating Barriers

When it comes to digging, many gardeners have found success by laying ½-inch (1.3 cm) hardware cloth over areas that seem to be squirrel targets. Cutting this wire fabric to fit and placing it over potted plants creates an effective barrier that impedes digging paws. However, this method is only effective against squirrels that feed above ground. If tunneling trespassers, such as ground squirrels, have invaded your yard with their burrow systems, the only sure way to protect garden beds is to line them with hardware cloth.

If small shrubs, tomato plants, berry bushes, or other established plants are the target of tenacious teeth, wire mesh cages are the answer. These can also be constructed out of hardware cloth or any weave of wire mesh in which the openings are no bigger than ½ by 1 inch (1.3 x 2.5 cm). A determined squirrel can squeeze into spaces you would never imagine. Cut panels to the size of the plant to be protected, then wire the panels together, or clamp them together with special clips sold at hardware stores.

Larger shrubs or trees often suffer damage from squirrels nibbling at the buds, collecting fruits, and stripping bark. Put a stop to this assault by attaching a flange to the base of individual trees as suggested on page 124, or by encircling the bark a few feet up from the base of the tree with a strip of 18-inch-wide (45.7 cm) metal flashing. The squirrels will scamper up as far as the metal, but they won't be able to get a grip on the slippery surface. Of course, take care to shield each tree to which squirrels have access, or they will simply leap from one to another. Protect young, tender bark with tree guard strips; you can either make these at home out of foil or plastic or purchase a commercial product. Wrap the strips around susceptible limbs, overlapping each wrap. Be sure to check the wrap job periodically for loosening or damage, and to allow for adjustments for the tree's growth.

**Squirrelly Facts**

The male tree squirrel takes twice as long to groom itself as the female does.

Since many squirrels can't seem to restrain themselves from going after freshly planted or sprouting seeds, it is wise to cover your garden in early spring with floating row covers. Be sure to anchor them around the edges with stones or other heavy objects to prevent squirrels from wriggling underneath. These flimsy covers can't stop a determined squirrel, but their mere presence stops most from investigating. The fabric has the added benefit of keeping away seed-stealing birds and flying insects; it also helps warm the soil in early spring for improved seed germination. Be sure to gradually loosen the fabric to allow for the growth of plants beneath, and remove it after the plants are well established.

## Fenced In (or Out)

You can keep ground squirrels out of your garden completely by surrounding it with a wire-mesh fence. The only drawback is that although a 2-foot-high (61 cm) fence will stop squirrels from going over it, the fence must also extend about a foot underground. A gas-powered trench digger is the perfect piece of equipment for excavating a long, narrow

Surrounding your garden with a recessed wire fence keeps even burrowing squirrels from intruding.

ditch around any area that needs to be fenced in this way. Available at most hardware stores, it does the dirty work and leaves a nice, neat trough. Roll out the fencing material, arrange it inside the ditch, and then attach the aboveground portion to stakes or posts.

## Repel by Smell

Some gardeners report success, at least temporarily, by using scent-based area repellents. Squirrels have a strong sense of smell and will detect the odor of a predator or some other unpleasant scent immediately. The catch is that they will eventually learn to ignore it. But in the ever evolving arsenal against squirrel encroachment, any weapon that works, even temporarily, is worth serious consideration.

Of the scent deterrents that have consistently been reported to have offended this most delicate of squirrel senses, blood meal is one of the best. This organic compound adds nitrogen and other valuable nutrients to your garden soil, to boot. Work the dry meal into the soil and water lightly. The aroma is heightened upon moistening, but most people don't find it objectionable. However, it apparently makes squirrels squeamish: They tend to avoid digging in blood-meal-amended soils.

As I mentioned in chapter 5, there are a variety of other area repellents marketed for use against squirrels, but most are ineffective in the outdoors. (See page 125 later in this chapter for more on repellents.) Even in the confines of ground squirrel tunnels, most repellents are ignored — the squirrels simply seal off offensive-smelling areas and dig around them.

## Not-so-tasteful Treatments

You would be hard pressed to find a taste that some squirrel, somewhere, doesn't crave. I've had ground squirrels eat my garlic from the tunnel up. But the prevailing wisdom is that squirrels don't care for anything that is hot enough to burn their mouths. Products based on capsaicin (see page 98), the "hot" part of hot peppers, have been around a while and have earned some degree of respectability. Try treating seeds, stems, and flower bulbs with either commercial or homemade hot pepper spray to ward off unwanted nibbling.

One concern about the use of hot pepper concoctions is that squirrels tend to grab their food with their front paws and then rub their eyes with those same paws. Some folks swear to having seen squirrels blinded by the use of hot pepper sprays.

Other products that can be used to treat plants include Ropel and Hinder. Both will leave a very bad taste in any squirrel's mouth. Soak bulbs before planting. Apply the chemical to ornamentals or to the stems of susceptible plants, but be prepared to reapply the product about once a month. While

### Keeping the Critters Out of the Corn

Dried hot peppers, ground and mixed with mineral oil and then painted onto the silks of corn ears, is an old timer's method for protecting corn crops.

Hinder is registered for use on food crops, unless you have a really tough palate or a strange appetite, *never* apply Ropel to any food grown for human consumption. Seed-treatment repellents that contain methiocarb (Mesurol) or thiram are useful against squirrels that go for freshly planted or just-sprouted seeds. Thiram in an 8 percent solution can be applied to plants to ward off a squirrel's toothy advances, but realize that it must be reapplied regularly to protect tender new growth as it emerges (a favorite treat of squirrels). Also, it cannot be applied to plants meant for consumption by either humans or livestock.

Consider not replanting, at least not for a season or two, whatever plants seem to hold the biggest attraction for your local squirrels. Once the squirrels learn that the goodies are not present, they may alter their range enough to exclude your tomato patch. Another possibility is that, given a couple of seasons, the guilty squirrels could expire, leaving young squirrels in their wake that never learned about your tasty crops.

### Not on the Menu

Hard though it may be to conceive, there are some plants that squirrels don't like. The following plants are largely ignored by foraging squirrels:

- Brussels sprouts, broccoli
- Castor oil plant
- Daffodil
- Foxglove
- Gopher spurge
- Hot pepper
- Monkshood
- Morning glory
- Squill

## Time Your Planting

For those nasty garlic-eating, stem-devouring, nonstop-digging, constantly reproducing ground squirrels that have dug their way into your domain, there is one very simple solution that takes advantage of the squirrel's natural tendencies. Grow plants that you can start at the end of winter or the beginning of spring. Early plantings often escape any ground squirrel damage, because the rodents are still asleep in their dens.

## Unleash the Hounds

Turning your dog loose on squirrels is a bittersweet deed. Watch those squirrels scamper! Yahoo. Look at 'em go. Darn dog. Get out of the azaleas, ya fool!

The problem with relying on your pooch to patrol your pansy patch is that most dogs, God love 'em, don't have much sense when it comes to which plants to trample and which to tiptoe around. In fact, Fido often gets just as much enjoyment out of digging and burying things in your yard as the squirrels seem to. Effectiveness depends entirely on the individual dog and the amount of time and talent you have to teach your pet what is acceptable and what isn't.

In general, some dogs are better at deterring ground squirrels and tree squirrels than others. The best dogs I've ever owned have been Dobermans. They're neat, intense, dedicated, beautiful, instinctively loyal, and able to chase a squirrel back up its tree in a single bound. Folks who hunt squirrels as game will tell you a good cur or feist of terrier bloodlines is hard to beat. Terriers will go nose to nose with tunneling squirrels in minutes, digging fast and furiously. And therein lies the rub — they dig bigger holes than the squirrels do. The upside, however, is that once ground squirrels realize there is a bigger, feistier "squirrel" in town, they will likely relocate. Nobody likes a dog tearing up their place.

Though it may dig bigger holes than the squirrel, a dog can help to deter unwanted pests.

Other critters that can police your plot include house cats, ferrets, and several varieties of snakes, including rat snakes, king snakes, gopher snakes, and boas, each of which can be deadly to squirrels. The scent of these predators alone may be enough to discourage the presence of squirrels in the areas they defend. Fake predators, from large plastic replicas of raptors to long, winding pseudosnakes, can be placed at strategic spots in the yard to startle and ward off squirrels. The clever rodents will quickly become accustomed to these pretend predators, so consider them only a temporary solution, and plan to move them routinely to keep the squirrels wondering.

## *Keeping 'em Outta the House*

It's hard to believe how much damage, both structurally and financially, these small bundles of fur can inflict, but believe it: Squirrels can wreak tremendous havoc in your house. It pays to keep them out.

### Limit Access

The most successful road to a squirrel-free house is to prevent them from gaining access in the first place. Odds are any squirrel that makes it *onto* the roof will also make it *into* the attic. It takes only a tiny crevice for squirrels to squeeze in, as they are flexible, determined trespassers and masters at remodeling even the most modest opening into a grand entrance. Follow these simple tips to prevent squirrels from setting paw on your roof:

- Make sure any branches that hang over your roof are trimmed back. Any tree limb within 20 feet (6.1 m) of your roof is an invitation to squirrels to drop in.
- If there are trees within leaping distance of your roof, consider attaching a 2- to 3-foot-wide (61–91.5 cm) metal collar around the trunk of each one, about 6 feet (1.8 m) up. To prevent damage to the tree, you might

want to consider affixing springs to the wire to allow for future growth. The metal collar works great in a yard that has only one or two trees, but where there are several trees, look to see if squirrels can bypass the collar by jumping from other perches, possibly even from trees on a neighbor's property.

- Install gutter guards and place screening over downspouts to prevent squirrels from using them to climb up to the roof.

## Repel Varmints

Squirrels turn to attics and other man-made structures as nesting sites for two reasons. One is that there may already be too many squirrels for the natural habitat to support and they don't have anywhere else to go, short of perilous migration. The other is — heck, it's there, it's comfy, it's downright inviting. You can change that.

Three types of repellents are useful in deterring squirrels from moving indoors: area or "stinky" repellents, such as naphthalene-based Rid-A-Critter; awful-tasting stuff, such as Ropel or Hinder; and contact repellents, such as 4-The-Birds and other sticky, irritating substances that squirrels hate to get on their fur. Placed in areas where squirrels can't avoid smelling, tasting, or touching, these products are effective in helping to keep the squirrels at bay. Depending on where and how they are used, some last longer than others. Read label directions carefully to avoid either wasting your money by buying too much or wasting your time by buying too little. One hundred percent naphthalene flakes or granules will repel squirrels when applied at a rate of 5 pounds (2.3 kg) per 2,000 square feet (185.8 square km) in an enclosed space. Be sure to stop up drafts, though, to prevent ventilation from carrying away the fumes. Apply at and near any potential points of entry, along known runs, and in nest sites for best results. Realize that if used alone, most repellents are short-term solutions at best. Plan to incorporate them into your overall plan.

## Blast Them Out

Another tactic that has met with mixed results is the ultrasonic wave machine. As discussed in chapter 5 (see page 114), the better units do have a practical application in home environments, as long as they are placed so that the waves are not displaced or absorbed by obstructions. In the attic, garage, or outbuilding, the best way to use them is to place them in a corner and aim them at the most likely point of entry.

The theory behind these units is sound, and the devices have proven effective. Failures are usually attributed to one or more of three things: incorrect use of the product, buying an inadequate product, or insufficient coverage that results from expecting one teensy unit to cover too large an area. To be effective against squirrels, the unit must have a setting of 30 kHz to 40 kHz. The lower setting is enough to give them a headache and a vaguely uncomfortable feeling, while the higher range is intolerable. These sounds are thought to mimic cries of distress, pain, warning, and possibly aggression. Remember to look for units that do not have paper speakers. Mylar is better, metal is best.

Ultrasonic wave machines mimic a squirrel's cry of distress and other repulsive sounds.

It is worth noting that although most people cannot hear within the range of these devices (normal human hearing falls somewhere between 3 kHz and 17 kHz), some can. Babies often hear well within this range. However, since you will be placing the unit in an attic or some other generally unused area, the sound waves will be blocked or absorbed by wood flooring, insulation, and other materials and shouldn't bother anyone.

Devices you have around the house may also be effective. Squirrels look for quiet, secluded, secure places to raise their young, which is one reason your attic is so appealing to them. Leave a radio tuned to an obnoxious station, or install timed or motion-sensitive floodlights. If the lights don't keep squirrels from moving in, at least they'll give your neighbors something to wonder about. For added protection, get a cat or ferret, and give it the run of the attic for a while every day. Your pet will love it, squirrels will avoid the place, and other wildlife will beat a hasty retreat as well.

## *Hello? . . . Anybody Here?*

As I discussed above, the most straightforward way to maintain a squirrel-free house is to seal the varmints out before they have a chance to barge in. But before you take the necessary steps to seal them out, make absolutely certain squirrels haven't already made your house their home. The only thing worse than a house full of squirrels is a house full of squirrels that can't get out.

### Searching for Clues

Squirrels have only two ways of getting into your house, or other buildings, from the outside. One is through holes that already exist, the other is by making holes. How can you tell if squirrels have already moved in and set up housekeeping? The first step is to find any potential points of entry.

Start at the top of your house and make a visual assessment of places squirrel burglars could break in. A walk on the roof is not out of order. Then work your way down, looking for the following:

- Open chimney pipe
- Loose shingles
- Holes in roof
- Holes in or under eaves (birds sometimes create these)
- Nests or other evidence in eaves troughs
- Gaps around, or loose-fitting, vents
- Accessible openings at fans
- Gaps or loose spots in siding
- Gaps or rotten, loose spots in fascia boards
- Foundation vents without covers
- Holes in or near foundation

Once you have found a potential point of entry, you need to determine if the squirrels have found it, too. A trick that would make Sherlock Holmes proud is to sprinkle flour, cornstarch, or baby powder at the hole. A fine mist of water or rubbing alcohol, sprayed before dusting the powder, will help it stick. (Obviously, this won't work during windy or rainy weather.) Come back in a day or two and check for tracks. Alternately, stuff the hole with wadded-up paper and wait to see if a perturbed squirrel tries to move it.

Unfortunately, you will not be able to find every tiny hole in your house that may allow a squirrel access. As a result, you will also need to search the house for signs of squirrel activity.

### *Look*

Climb into the attic and take a good look around. Squirrels will unwittingly make their presence known in the form of pellets, chewed- and fluffed-up insulation, and the remnants of nuts or pinecones.

### *Sniff*

Squirrels are often detectible by the little presents they leave. An enclosed, poorly ventilated area, such as an attic, soon acquires a characteristic odor if animals are using it as a toilet.

### *Shine*

Forget the flashlight — grab a black light, instead. Believe it or not, squirrel urine glows under black light, making it very easy to detect.

### *Listen*

Squirrels have no need to be quiet. In fact, as their calls and chatter indicate, they are naturally noisy little rabble-rousers. Also, their romping movements can echo through-out a house, especially at night, when they sound like small elephants in the walls or the ceiling. Tree squirrels, thank-fully, make a ruckus during the day; it's the flying squirrels that keep you up nights with their commotion.

### *Look Again*

If you believe you have squirrels in residence, determine where the nest is and serve them an eviction notice. (See below.) Sometimes, it helps to have a cat or dog assist you in ferreting out nest sites. Come to think of it, a ferret — closely related to such natural enemies of squirrels as the marten, weasel, and mink — would work just fine, too.

Remember: In the wild, individual squirrels often main-tain several different nests, so don't expect any less of them in your attic. Be thorough in your search for nest sites, and don't stop looking just because you find one.

## Evicting Squirrels

Once squirrels have made themselves at home in your home, you will have a tough time convincing them to leave. Repellents, ultrasonic noise devices, glue boards, and traps meant for rats and mice have all been tried, but resident squirrels are very territorial. Once squirrels have staked a claim, about the only realistic way to get rid of them is to have a professional remove them (see page 136).

## An Ounce of Prevention . . .

After you've gone to the trouble of removing stowaway squirrels, don't make the mistake of believing your home is safe. A great deal of a squirrel's life is governed by its sense of smell. In their routine comings and goings through whatever openings they have found into your attic, squirrels leave scent trails that could beckon to other squirrels to follow. The scent is detectable by other squirrels for several months.

Begin by thoroughly scrubbing known entrances with an ammonia-based cleanser. Swab down any known immediate pathways to these entrances as well, such as a favorite route along the window trim. This will cut the scent and reduce the chances that other squirrels will follow. Also, scrub out or douse any nest sites using a spray bottle of diluted ammonia to further erase the smelly claim squirrels have made.

As a final measure, apply area repellents, such as Rid-A-Critter, in the old nests to discourage new squirrels from checking out the premises. Mothballs (the fresher, the better), [methyl nonyl ketone] crystals, and paradichlorobenzene (moth crystals) are also effective at temporarily repelling squirrels in enclosed spaces. Another substance that squirrels detest is 4-The-Birds, a tacky, irritating gel that sticks to them when they walk across it. Spread some around entrances or along any likely paths they may have taken to get in, such as poles, heavy wires, or other obvious routes. It is available

### Drop Them a Line

If you suspect that a squirrel has fallen down the chimney and can't get out, it can't hurt to offer a helping hand – or rope. The Society for the Prevention of Cruelty to Animals (SPCA) recommends lowering a rope or length of knotted sheet down the chimney for the animal to climb back out. Dust the chimney top with baby powder to monitor whether the trapped critter has indeed escaped, then cover the chimney with a cap or guard.

in a liquid form for application by pump sprayer over large areas that can't be sealed off by other means.

## Never Say Never

Clever and persistent, squirrels seem to enjoy solving problems, such as how to get to that tomato plant, fruit tree, or bird feeder. Consider using several different squirrel-repelling tactics, and rotate them on a random basis.

Never attempt to seal a squirrel or any wild animal out of its established nesting site without relocating it elsewhere or humanely destroying it (which is often more humane than relocation). Sealing off entry holes can be traumatic for any squirrel accidentally trapped inside the house and, needless to say, disastrous for your attic as the frantic critter tries to chew its way back out. Allow about two weeks after eviction to be sure all squirrels are *really* gone, and then proceed with sealing off any and all openings.

Time your repairs to avoid those periods when squirrels are most likely to be inside or, worse, to have young nesting inside. All but the nocturnal flying squirrels are most active from about two hours after sunrise to about two hours before sunset. If you suspect that tree squirrels are trying to move in, make repairs during the day when they are most likely to be out and about. Avoid very hot days (they retreat to their nests to snooze) and windy, stormy days (also off times for squirrels). Also, realize that most squirrels raise one litter of young in spring, usually between March and May. If possible, arrange to make your repairs either before the squirrels are most likely to have babies in the nest or long after they have left.

Use concrete to patch up holes or cracks in the foundation, and use sheet metal or hardware cloth ½ by 1 inch (1.3 x 2.5 cm) for openings beneath eaves. Be sure to replace broken or missing shingles, taking care to repair any structural roof damage first. Nailing a shingle over a hole won't solve any problems. Install chimney caps or specially designed screens on every chimney pipe. Caps are relatively inexpensive and have the added benefit of protecting your roof from chimney sparks and keeping snow and rain from falling down it. Seal the gaps around vents by screwing the edges down flush and applying weatherproof sealant around the edges. Repair any loose or rotten spots in the siding or fascia. Attend to foundation vents. Replace any plastic vents with metal ones. If necessary, crawl under the house and fit grates or hardware cloth on the insides to prevent rodents from getting through. Check for any cracks or holes in the foundation and have them filled. Finally, scan the entire perimeter of the house for other points of entry and block them as necessary.

## *Squirrels on the Wire*

Just when you thought it was safe to come out of the attic, squirrels have found another main highway leading straight for it. Before you know it, they are right back in your house. Where are they coming from now? Take a look at the wires around your house. They may be the source.

### Wired for Trouble

Overhead utility wires are favorite routes for squirrels for obvious reasons. They aren't exactly crowded thoroughfares — squirrels always have the right-of-way, even when birds are perched along the lines. They are at a comfortable height (for squirrels, that is), well above the riffraff below. And overhead utility lines almost always go someplace interesting: your attic, say, or maybe that nifty little area over the garage you were planning to fix up.

Squirrels on utility lines are more than just a nuisance to homeowners. Millions of dollars in damage to public utilities is attributed each year both to squirrels chewing on them and to squirrels coming in contact with transformers and blowing themselves up. In 1997, it cost the telephone company Bell Atlantic more than $300,000 to install a polyvinyl-chloride shield over many of their phone lines in squirrel-plagued Richmond, Virginia, just to spare the squirrels and their customers from problems with their phone service. Considering that this company opted for this expenditure to *save* money, think about how expensive a problem the squirrels must have been.

**Squirrelly Facts**

Squirrels are very good swimmers. Migrating squirrels moving from one side of a river to the other may pass their brethren vigorously squirrel-paddling in the opposite direction.

Elsewhere, the *Bismarck Tribune* reported that an estimated 15 percent of power outages in the Montana–Dakota utility systems were caused by squirrels zapping themselves on transformers. Plastic jackets had been installed over the transformers, but over time they had begun to lose their effectiveness, and outages were on the rise. Enter a device invented by Jim Guthrie of Iowa City called, fittingly enough, the Guthrie Guard, which sits atop the transformers. Made from stainless steel and polypropylene, the devices are mounted with short spikes that give off a warning jolt to any squirrel that comes too close. About 2,500 units were installed by mid-1997. The result? No more fried squirrels, money saved on man-hours to repair lines, and happy customers.

## No More Walking the Tightrope

If squirrels are a high-wire nuisance in your area, here are a few suggestions you may want to try:

- Cover any utility wires or other aerial means of access with 2-foot-long (61 cm) sections of 2- to 3-inch-diameter (5.1–7.6 cm) plastic or PVC pipe. Cut a long, narrow slit down the length of each section, spread the pipe open, and fit it over the wire. The pipe sections act as tumblers should squirrels attempt to cross.

- Coat the ends of the wire that the squirrels are most likely to jump onto with 4-The-Birds or some other tacky or greasy substance.
- Install bafflelike shields at the accessible ends of the wire to serve as a roadblock. A 2- to 3-foot-wide (61–91.5 cm) circle of metal or acrylic, with a hole cut in the center and a slit from center to the outside, can easily be slipped over utility wires.
- Always call your local utility before attempting to work on overhead lines! You, like the squirrels, could get zapped!

## *If All Else Fails . . .*

So you've tried every preventive technique in the book and the squirrels are still driving you nutty. Enough is enough! You are at wit's end and desperate for a solution. If you can't stomach losing to the wily varmints, consider trapping as a last resort.

About the only time that trapping squirrels is a viable solution is when they have invaded your house or attic to the point where you have no recourse but to permanently remove the offenders. Trapping is not an effective way to manage squirrel populations.

### Some Background

If you embark upon trapping as a means of squirrel control, you are responsible for knowing what you are doing. Here's the skinny:

- Trapping does not effectively reduce squirrel numbers in your area. For every Tom Squirrel you remove, Dick and Harry Squirrel are just waiting to take his place. Given that roughly half of the existing squirrel population expires each year, how do you know the squirrel you nabbed wasn't on its way out anyway?

- Sorry, but live-trapping is a sham. Once you've caught a squirrel, what do you do with it? You can't sell it to the circus. The standard advice, depending on the source, recommends releasing it at least 3, 5, or 7 miles (4.8, 8.1, or 11.3 km) from your home. But this is to prevent the squirrel from returning, and possibly to ease your conscience. Heaven knows, it does the squirrel little or no good. Some squirrels are so traumatized by the capture and handling that they perish from shock. For those that survive, being released into strange territory either traumatizes the squirrel or upsets the natural balance of the new location. Trapping one squirrel may be killing several. The SPCA warns against trapping and relocating squirrels from midspring through early autumn; this is when they are most likely to be nursing young.

## Put Them on the Pill

British scientists are toying with a way to slow down the rate of squirrel reproduction. In other words, they are trying to come up with birth control for squirrels. Sheffield University in northern England has launched a program to test a protein vaccine designed to make female gray squirrels infertile. The squirrels take the "pill" in the form of bait, which could be left in the woods, near bird feeders, or at squirrel-feeding stations. Over time, this could reduce the current population of the pesky intruders significantly.

- Trapping in some regions is regulated, and a permit from your state department of natural resources or game department may be required for catching squirrels outside regularly scheduled hunting seasons. In some areas, a re-release permit is also required. Realize that in many areas, trapping involves more than just you and the squirrels. In Illinois, for instance, a state conservation officer or wildlife biologist must first confirm that there is a wildlife problem (squirrel in attic) and that there is no alternative solution to trapping (squirrel won't leave no matter how many times you ask). The upside is that you may be able to call an animal-control officer to do the dirty work for you.

- Realize that there are professional pest-control companies that are up on all the regulations and happy to visit your attic to remove any unwanted critters. Fees vary, so call several and ask for estimates. Some will even come to your home and offer a free estimate and dispense a little advice to boot. Be ready for some horror stories!

## Trapping Tips

If you insist on trying to trap the squirrel yourself, use a live trap such as the classic Havahart traps. Havahart traps may fail because squirrels can roll them over and cause the door to open, or wedge their tails beneath the door when inside, preventing it from falling shut and trapping them. Be sure to pick the type and size of trap appropriate for the squirrels you are after. Some further hints:

Release a trapped squirrel 3 to 7 miles (4.8–11.3 km) from your home to prevent it from returning. Note that captured and relocated squirrels may die from shock.

- Place traps along known squirrel runs, near bird feeders, along fence lines, and so on. Don't place traps near nests; this makes the squirrels suspicious.
- Always place traps on a firm surface. Never hang them from trees or other objects.
- Bait traps with goodies such as shelled pecans, peanuts, peanut butter, sunflower seeds, or apple or orange slices. You can also use a professional product, such as Trappers Choice Pecan Paste.
- For flying squirrels, place traps where squirrels are known to be active (such as in the attic) and bait with pecan paste and sunflower seeds.
- Check traps only once a day to avoid tipping the squirrels off as to what you're up to.
- Place most of the bait behind the trip pan of the trap, with a little on the pan itself.

CHAPTER 7

# Squirrel-proof by Design

**In most cases** the best way to coexist with squirrels of any kind is to plan ahead; try and outsmart them before they cause any problems. Remember, most of them have a brain somewhere along the size of a walnut. We can do this.

## *Squirrel-proof Bird Feeders and Nestboxes*

Squirrel-proof bird feeders and houses are available commercially in a variety of designs. Similar devices can be manufactured at home for far less than commercial costs of $30 and up. Another option is to customize existing feeders and nestboxes with products or add-ons designed to keep squirrels at bay.

## Acrylic Tube Feeder

This squirrel-proof feeder is adapted from a popular, inexpensive acrylic tube feeder. Spring loading the perches allows birds to feed undisturbed but pitches heavier squirrels into the air or onto the ground. Hang this one near a window and chuckle at the thwarted squirrels.

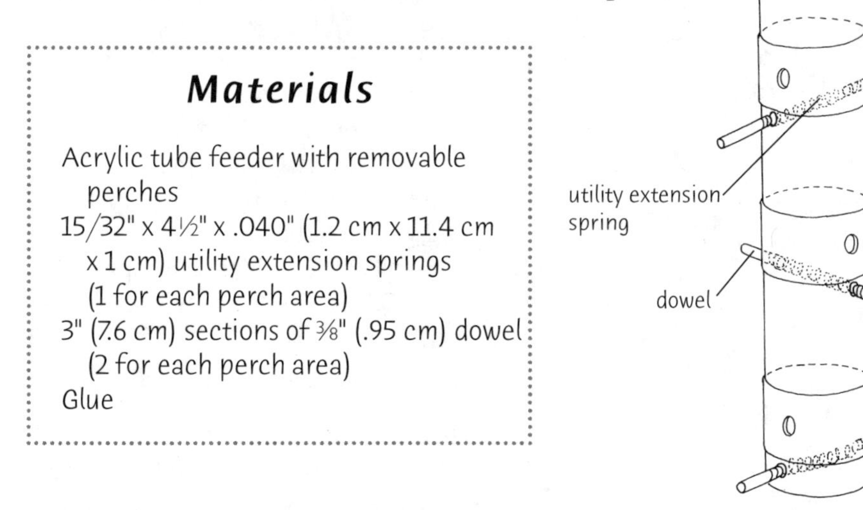


### *Materials*

- Acrylic tube feeder with removable perches
- 15/32" x 4½" x .040" (1.2 cm x 11.4 cm x 1 cm) utility extension springs (1 for each perch area)
- 3" (7.6 cm) sections of ⅜" (.95 cm) dowel (2 for each perch area)
- Glue

1. Remove the perches and ends from the tube feeder.
2. With your hand inside the tube, feed a spring through each set of perch holes.
3. Put glue on one end of each dowel and squeeze these ends inside the ends of the springs.
4. Let the glue dry, then tug on the new spring-loaded perches to ensure that the dowels and springs are fastened securely.
5. Replace the bottom, then fill with seed and hang.

## Recycled Spring-loaded Feeder

Simultaneously stymie squirrels and recycle household materials. This primitive but effective feeder has a spring-loaded barrier that covers the feeder holes when a squirrel jumps on the perch.

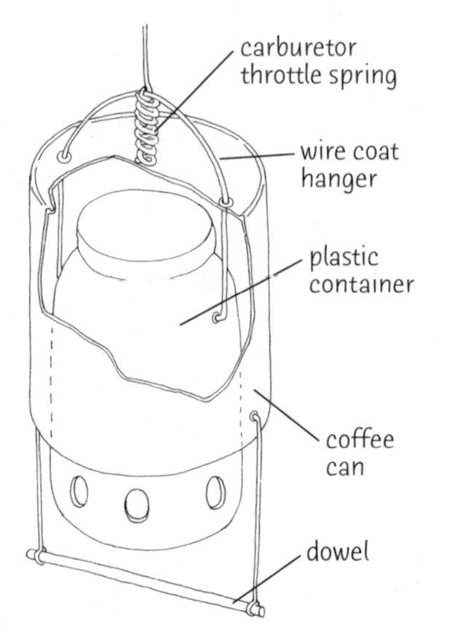


### Materials

64 oz. (1.9 L) mayonnaise plastic container (or any like-sized plastic container)
39 oz. (1.2 L) coffee can
2 wire coat hangers
Small carburetor throttle spring
7" (17.8 cm) length ⅜" (.95 cm) wooden dowel

1. Drill or punch the following holes in the plastic container: three ½-inch (1.3 cm) feeding holes near the bottom of the container and two small holes opposite each other near the top of the container.
2. Drill or punch the following holes in the aluminum can: two small holes opposite each other in the bottom of the can, one small hole in the bottom center, and two small holes opposite each other near the top of the can.
3. Straighten one hanger and cut one 12-inch (30.5 cm) section and two 4-inch (10.2 cm) sections. Bend the long section into a U shape.
4. Use pliers to bend one end of the U wire into a hook. Insert the hooked end into one of the holes near the top of the plastic container.
5. Feed the free end of the wire through the two holes in the bottom of the aluminum can. Bend the end of the wire as in step 4 and insert the end into the opposite hole in the plastic container.

6. Use pliers to bend the ends of the two 4-inch (10.2 cm) wires around the ends of the dowel. Bend the free ends of the wires into hooks as in step 4. Insert these ends into the holes near the top of the aluminum can to hang the perch.
7. Slip one end of the spring into the hole in the bottom center of the can. Slip the other end over the hanger. Use pliers to bend the ends of the springs so that it is fastened securely.
8. Fill the plastic container with seed and hang using other straightened coat hanger.

### Cube-in-a-cube Feeder

Small birds such as titmice and chickadees can enter the cube and feed, but the mesh prevents the entry of squirrels. The 4-inch (10.2 cm) gap around the seed also keeps their criminal limbs from snagging illegal snacks.

#### *Materials*

- 24" x 48" (61 cm x 1.2 m) piece of 1¼" (3.2 cm) mesh wire
- Board or hardcover book
- Masking tape or a marker
- 8" x 16" (20.3 cm x 40.6 cm) piece of ¼" (.63 cm) mesh wire
- 4 feet (1.2 m) of 24-gauge wire
- Four 4" (10.2 cm) S hooks
- Eight ¾" (1.9 cm) retaining rings
- 12" x 12" (30.5 cm x 30.5 cm) piece of window screen

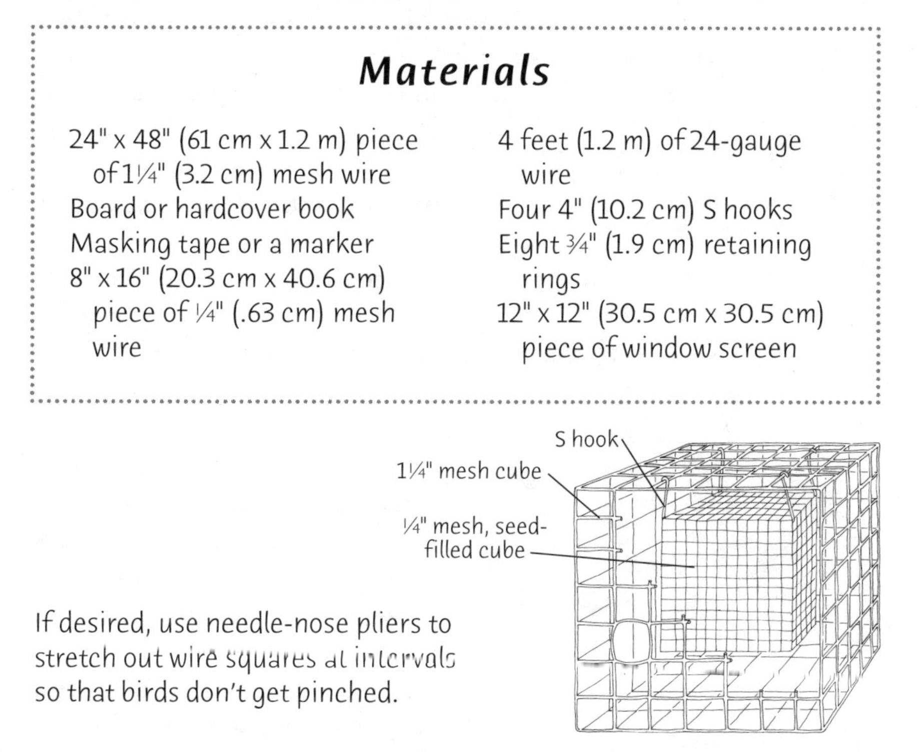


If desired, use needle-nose pliers to stretch out wire squares at intervals so that birds don't get pinched.

*Note: Larger mesh will allow more bird varieties to reach the food but might also permit the entrance of small squirrels.*

1. Wearing protective gloves, use wire snips to cut out the 24" x 48" (61 cm x 1.2 m) L shape shown in the diagram below from 1¼-inch (3.2 cm) mesh. Using tape or a marker, place a mark every 12 inches (30.5 cm) on the long side of the L.
2. Using a board or hardcover book as a guide, fold the bottom segment of the L 90°. Next, fold the shape at each 12-inch (30.5 cm) marker to form a topless cube. Use 2-inch (5.1 cm) lengths of wire to secure the cube on the bottom and side.
3. Cut out the 8" x 16" (20.3 x 40.6 cm) L shape from ¼-inch (.63 cm) mesh. Using tape or a marker, place a mark every 4 inches (10.2 cm) on the long side of the L.
4. As in step 2, fold the bottom segment of the L 90°. Next, fold the shape at each 4-inch (10.2 cm) marker to form a topless cube. Wire the cube closed on the bottom and side.
5. Fill the small cube with seed that won't slip through the mesh, such as sunflower. Next, cut a 4-inch (10.2 cm) square from the ¼-inch (.63 cm) mesh. Wire this piece to the top of the seed-filled cube and attach the four S hooks near its corners. (As an alternative to wiring on the top, since you will often remove and replace this piece, attach the top using four retaining rings.)
6. Place the 12" x 12" (30.5 cm x 30.5 cm) piece of window screen in the bottom of the large cube to catch spilled seed and allow rainwater to drain.
7. Cut a 12-inch (30.5 cm) square from the 1¼-inch (3.2 cm) mesh. Attach the hooks of the small seed-filled cube to this piece. Place the suspended small cube inside the large cube and wire the top of the large cube closed. (As an alternative to wiring on the top, use four retaining rings as in step 5.)
8. Use remaining wire to hang.

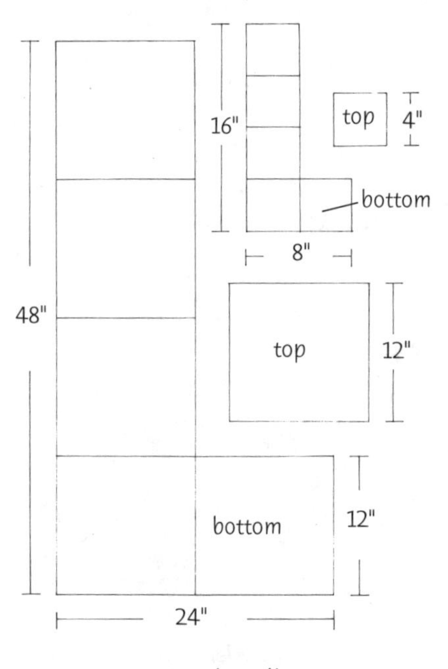

Cutting diagram

## Pivotal Pine Feeder

If you are a bird enthusiast who is both ambitious and concerned with aesthetics, try this pine feeder that operates on a pivot system to keep out squirrels.

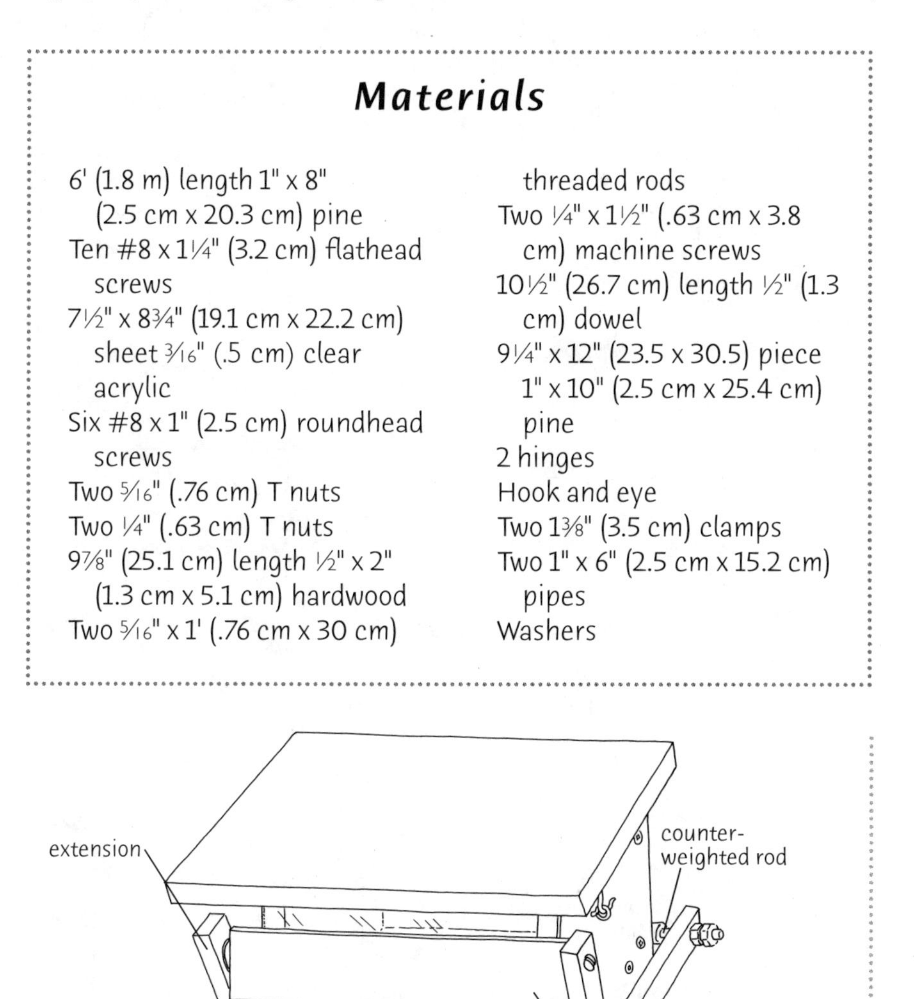

### *Materials*

- 6' (1.8 m) length 1" x 8" (2.5 cm x 20.3 cm) pine
- Ten #8 x 1¼" (3.2 cm) flathead screws
- 7½" x 8¾" (19.1 cm x 22.2 cm) sheet 3⁄16" (.5 cm) clear acrylic
- Six #8 x 1" (2.5 cm) roundhead screws
- Two 5⁄16" (.76 cm) T nuts
- Two ¼" (.63 cm) T nuts
- 9⅞" (25.1 cm) length ½" x 2" (1.3 cm x 5.1 cm) hardwood
- Two 5⁄16" x 1' (.76 cm x 30 cm) threaded rods
- Two ¼" x 1½" (.63 cm x 3.8 cm) machine screws
- 10½" (26.7 cm) length ½" (1.3 cm) dowel
- 9¼" x 12" (23.5 x 30.5) piece 1" x 10" (2.5 cm x 25.4 cm) pine
- 2 hinges
- Hook and eye
- Two 1⅜" (3.5 cm) clamps
- Two 1" x 6" (2.5 cm x 15.2 cm) pipes
- Washers

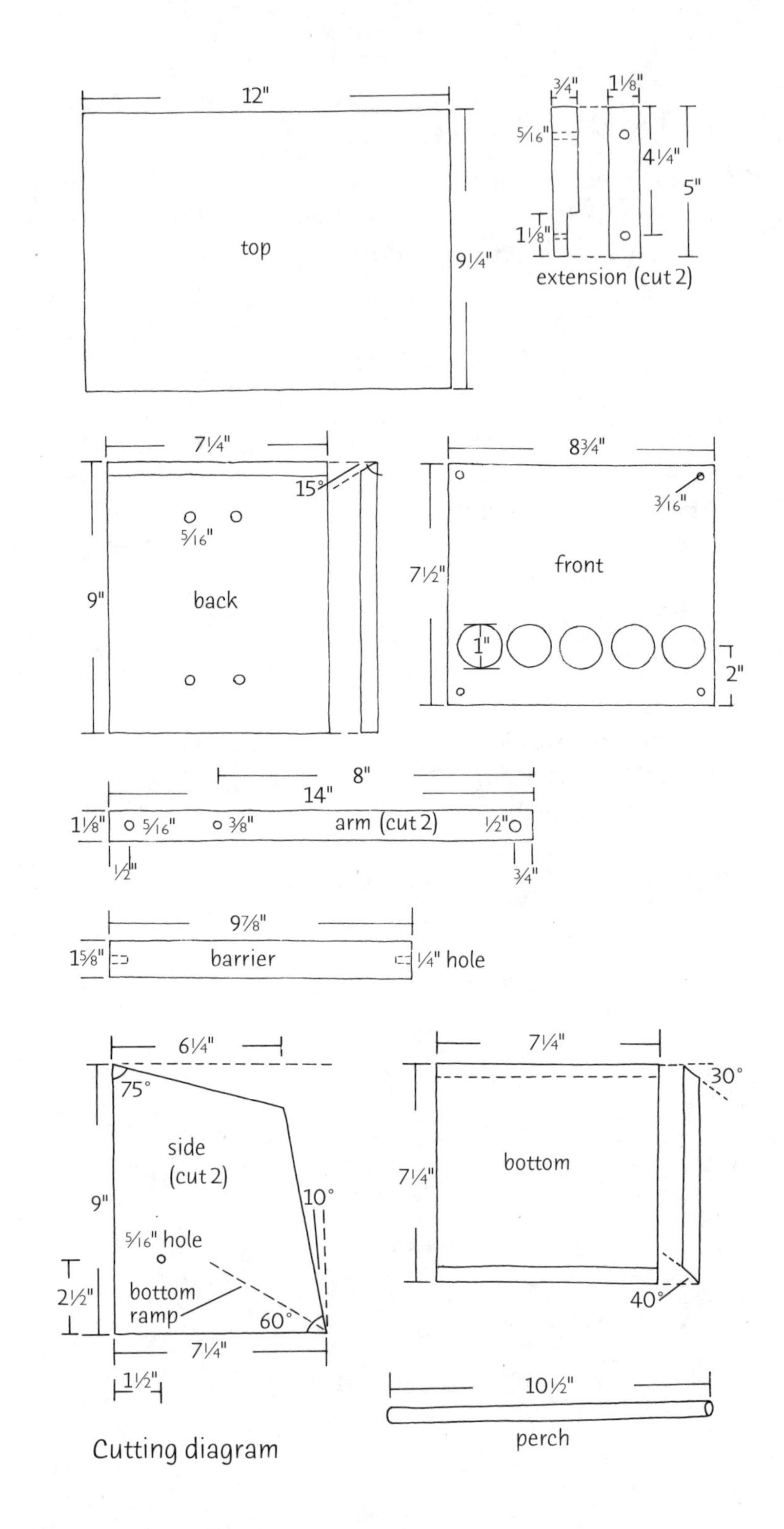

Cutting diagram

1. Use a handsaw or table saw to cut the back, bottom, and sides of the feeder from 1" x 8" (2.5 cm x 20.3 cm) pine.
2. Drill and countersink holes, then attach the sides to the back using #8 x 1¼" (3.2 cm) flathead screws. Attach the bottom to the sides at a 60° angle.
3. Drill a 5⁄16-inch (.76 cm) hole in each of the sides 2½ inches (6.4 cm) up from the bottom and 1½ inches (3.8 cm) in from the back.
4. Leaving on the protective coating, drill the acrylic sheet corners for four #8 x 1" (2.5 cm) screws and drill five 1-inch (2.5 cm) feeder holes. (A drill press is helpful for this step.) Screw the acrylic front to the feeder using roundhead screws.
5. Cut two arms as shown and drill two ½-inch (1.3 cm) holes for the perch, two ⅜-inch (.97 cm) holes for the pivot rod, and two 5⁄16-inch (.76 cm) holes for the weight rod. Place 5⁄16" (.76 cm) T nuts in arms at pivot.
6. Cut two extensions as shown and drill two 5⁄16-inch (.76 cm) holes for the T nuts and machine screws and two 1 1⁄64-inch (2.6 cm) holes for the roundhead screws. Place ¼" (.63 cm) T nuts at extensions.
7. Cut the barrier from hardwood and drill the centers of the ends with a ¼-inch (.63 cm) drill.
8. Assemble arms, pivot rod, and weight rod. Attach barrier to extensions using machine screws. Then, using roundhead screws, attach the extensions to the arms 3½ inches (8.9 cm) from the front of the arms. This is the key adjustment for enabling the barrier to swing up and down.
9. Dab glue in the perch holes in the arms and slide in the dowel.
10. Cut the feeder top from 1" x 10" (2.5 cm x 25.4 cm) pine. Attach the top to the back using hinges, then attach the hook and eye.
11. Slide the pipes onto the weight rod and use washers to add further weight if needed.
12. Drill four 5⁄16-inch (.76 cm) holes in the back as shown and attach the two clamps to the back for mounting.
13. Mount feeder on pipe or post and fill with seed.

## Squirrel-proof Bird Houses

Since a squirrel's second favorite thing after stealing from birds is to kill their babies, any bird lover worth his salt should consider squirrel-proof bird house designs. The trick here is simple — use something they can't chew through.

Squirrels most often raid bird houses by gnawing around the entry hole until it is large enough to squeeze through. A simple design that prevents them from gaining access is one that includes a metal strip or wooden block fastened around the opening.

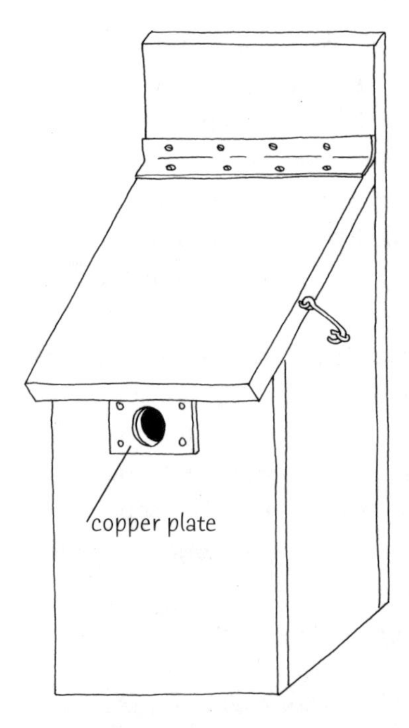


Even the most determined squirrels can't chew through metal. Fastening a copper plate to the entryway keeps squirrels from enlarging the hole and getting into your nestbox.

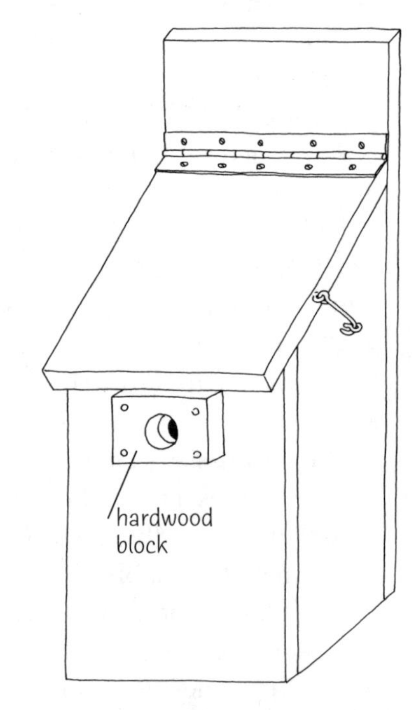


A hardwood block serves the same purpose, but it may eventually need replacement. Vigorously gnawing squirrels may whittle it away in their break-in attempts.

# Resources

## Scent Repellents

### Whatever Works Garden and Home Pest Control

Earth Science Building
74 20th Street
Brooklyn, NY 11232
(800) 499-6757
www.whateverworks.com

***Fox Urine, Coyote Urine:***
*Gives squirrels the illusion that predators are near.*

***Scoot Squirrel:***
*Spray designed to keep squirrels away from plants and bird feeders without harming animals or plants.*

***Squirrel Chaser Pouches:***
*Contain all natural ingredients that repel squirrels.*

**Rid-A-Critter**

*The fumes from these time-release granules drive away squirrels and rabbits.*
Dr. T's Nature Products
P.O. Box 682
Pelham, GA 31779
(800) 299-NATURE (6288)
www.animalrepellents.com
*Also available from U-SPRAY, Inc. (See below under Ropel.)*

## Taste Repellents

**Squirrel Away**

*Nutritional supplement for wild birds that is distasteful to squirrels and other mammals.*
Scrypton Systems, Inc.
P.O. Box 404
Annapolis, MD 21404
(800) 229-5454
info@scrypton.com
www.scrypton.com

**Ropel**

*An awful-tasting, odorless spray that discourages squirrels from chewing on bird houses, wires, etc.*
U-SPRAY, Inc.
4653 Highway 78
Lilburn, GA 30047
(800) 877-7290
Fax: (770) 985-9319
uspray@mindspring.com
www.bugspray.com

## Tactile Repellents

**Nixalite**

*This flexible, spiny barrier creates "off-limits" areas for squirrels.*
Nixalite of America, Inc.
1025 16th Avenue
East Moline, IL 61244
(888) 624-1189
nixalite@qconline.com
www.nixalite.com

**4-The-Birds**
*This tacky substance, available in gel or liquid form, keeps squirrels at bay by irritating their skin.*
U-SPRAY, INC
4653 Highway 78
Lilburn, GA 30047
(800) 877-7290
Fax: (770) 985-9319
uspray@mindspring.com
www.bugspray.com

## Baffles

**ERVA Squirrel Baffles**
*This line of squirrel baffles includes pole-mounted and hanging models.*
ERVA
1895 N. Milwaukee Ave.
Chicago, IL 60647l-4464
(800) 342-3782

### Wild Birds Forever
27202 Highway 189
P.O. Box 4904
Blue Jay, CA 92317-4909
(800) 459-BIRD
www.birdsforever.com

***Woodlink Pole Mount Squirrel Baffler:***
*You can clip this unique squirrel baffle around your pole without having to remove your bird feeder. Constructed of black, unbreakable textured powder-coated steel. 14" diameter by 5¼" high.*

***Droll Yankee Hanging Squirrel Guard:***
*Clear polycarbonate dome and adjustable brass rod for use over any hanging feeder. 15" diameter.*

***Hyde 20" Hanging Squirrel Baffle:***
*A large hanging baffle ideal for protecting larger feeders. Made of unbreakable polycarbonate. 20" diameter by 4" tall.*

## Ultrasonic Devices

**The Mole Chaser**
*This ultrasonic device is inserted into the ground; rodents can't tolerate the frequency it emits.*
Tower International
1829 Hubbard Lane
Grants Pass, OR 97526
(800) 955-8352
staff@molechaser.com
www.molechaser.com

**Transonic Mole/Gopher Repeller**
*Repels moles, gophers and other ground rodents with ultrasonic waves.*
Biocontrol Network
(615) 370-4301
info@biocnet.com
www.biconet.com

**Rodent Ultrasonic Repellent**
*Emits a high sound audible only to rodents, including squirrels.*
U-SPRAY, Inc.
4653 Highway 78
Lilburn, GA 30047
(800) 877-7290
Fax: (770) 985-9319
uspray@mindspring.com
www.bugspray.com

### Whatever Works Garden and Home Pest Control
Earth Science Building
74 20th Street
Brooklyn, NY 11232
(800) 49WORKS
(800-499-6757)
www.whateverworks.com

**Transonic 300:**
*Emits ultrasonic frequencies covering up to 3,000 square feet. Offers two sound levels.*
**Squirrel Deterrent:**
*Emits ultrasonic waves covering a 56' x 14' area.*

## Traps and Trapping Supplies

**Collapsible Squirrel Trap**
*Designed to catch small animals without harming them, indoors or out. Rustproof collapsible construction. Bait included.*
Deerbusters
9735 A Bethel Road
Frederick, MD 21702
(888) 422-3337
www.deerbusters.com

## Squirrel-proof Bird Feeders

**Lee Valley Tools Ltd.**
*Sells a variety of feeders impenetrable by squirrels and large nuisance birds.*
customerservice@leevalley.com
www.leevalley.com
U.S. orders:
P.O. Box 1780
Ogdensburg, NY 13669-6780
(800) 871-8158
Can. orders:
P.O. Box 6295, Stn. J
Ottawa, Ontario K2A1T4
(800) 267-8767

**Wild Birds Unlimited, Inc.**
*From the extravagant to the utilitarian, this national franchise carries a range of prefab, squirrel-proof bird feeders.*
Wild Birds Unlimited, Inc
11711 N. College Ave., Suite 146
Carmel, IN 46032
(888) 302-2473
webmaster@wbu.com
wbu.com

**Squirrel Beater**
*When a squirrel crawls down over the top of the feeder or lands on the outside cage, its weight pulls the cage down over the tube, closing off the feeding ports.*
Five Islands Products, Inc.
P.O. Box 2095
Scarborough, ME 04070-2095
(207) 885-1580
www.britesites.com

**Stainless Steel Nuttery Bird Feeder Q20**
*Squirrels and other small animals can't damage the stainless steel construction. Holds a pound of nuts or sunflower seeds. 5" diameter by 12" tall.*
Bird Watcher's Marketplace
3150 Plainfield NE
Grand Rapids, MI 49525
(800) 981-BIRD (2473)
order@birdwatchers.com
www.birdwatchers.com

# *Index*

Page references in *italics* indicate illustrations.

# *Other Storey Titles You Will Enjoy*

**Deer Proofing Your Yard & Garden,** by Rhonda Massingham Hart. Includes simple, effective, and natural strategies for repelling deer, including commercial and homemade deterrents and landscaping with deer repelling plants. 160 pages. Paperback. ISBN 0-88266-988-5.

**Bugs, Slugs & Other Thugs,** by Rhonda Massingham Hart. Includes hundreds of ways to stop pests without risk to the user or the environment, from folk remedies to the latest scientific discoveries. 224 pages. Paperback. ISBN 0-88266-664-9.

**Hand-feeding Backyard Birds,** by Hugh Wiberg. Following the full-color photographs and step-by-step instructions in this book, anyone can learn to hand-feed chickadees, nuthatches, titmice, and other wild birds. 160 pages. Paperback. ISBN 1-58017-181-8.

**Backyard Birdhouse Book,** by René and Christyna M. Laubach. Written by two experienced naturalists, this practical guide offers complete plans for eight easy-to-build bird houses tailored to the needs of 25 cavity-nesting species. 216 pages. Paperback. ISBN 1-58017-104-4.

**The Backyard Bird-Lover's Guide,** by Jan Mahnken. A gorgeously illustrated volume brimming with information about attracting, enjoying, and understanding 135 of North America's most common bird species. 320 pages. Paperback. ISBN 0-88266-927-3.